HARPOONER

A Four-Year Voyage on the Barque Kathleen

1880–1884

By
ROBERT FERGUSON

Edited by
LESLIE DALRYMPLE STAIR

Illustrated by
PAUL QUINN

UNIVERSITY OF PENNSYLVANIA PRESS: PHILADELPHIA: 1936

Copyright 1936
UNIVERSITY OF PENNSYLVANIA PRESS
Manufactured in the United States of America

LONDON
HUMPHREY MILFORD
OXFORD UNIVERSITY PRESS

DEDICATED TO
Samuel Hazzard
The Crew of the
Kathleen
and the
New Bedford Whaling Fleet

PREFACE

To every man worthy of the name comes the desire to see the world and feel the thrills of adventure. For those who from necessity are obliged to remain at home, this longing can be satisfied only by proxy, by reading or hearing the glamorous tales of those who have been so fortunate as to travel to far distant places.

Recently the editor was permitted to read the personal diaries of Robert Ferguson covering his sea voyages from 1875 to 1885. One of these, a whaling voyage on the barque *Kathleen* of New Bedford, of over four years' duration, from May 5, 1880, until May 22, 1884, proved so interesting and full of adventure that it was believed that others should share in its enjoyment.

No record of a whaling voyage, and especially that of one made by a New Bedford whaler before the whale-oil industry became unprofitable, could be more epochal and historic than that found in this diary of Robert Ferguson covering his experiences on shore and at sea while harpooner on the barque *Kathleen*. It is intimate in the details of his own life; his shipmates and superior officers on board; his friends on shore; and in the unveiling of his innermost thoughts.

One cannot fail to be impressed by his staunch character. He was fearless in the face of danger; uncomplaining throughout hardships; God-fearing among rough men; a fighter when necessary; yet always considerate to both superiors and inferiors in rank. His schooling ended when he was obliged to go to work at the age of nine years. When he was nearly seventeen years old he ran away to sea, but continued to study and better himself until he overcame the difficulties of navigation and became the master of a ship. Not content with improving his own education, Captain Ferguson continually helped those who had none.

It has been endeavored to retain the naive, succinct style of the original manuscript with its quaint colloquialisms and its maritime flavor. No additions have been made except those necessary for purposes of elucidation. Condensation has been done only when required to eliminate repetition during routine periods. As for deletions due to vulgar thought or language, none were necessary.

Captain Ferguson's diaries were originally written in miscellaneous notebooks. These notes were later copied into ledgers. This explains the occasional use of slang expressions which seem to be more modern.

The barque *Kathleen*, 312 tons, belonged to the same owners as the barque *Charles W. Morgan*, frequently referred to in this diary. The latter ship may be seen today at the estate of Colonel Green at South Dartmouth, Mass., where it has been permanently set up for the use of the public as a monument and record of a typical barque used in the whaling industry. The *Kathleen* was built in Philadelphia in 1844 for the merchant service. Soon after, she was bought in New York by Captain James Slocum and fitted for a whaler. In 1857, the barque was sold to J. & W. R. Wing & Co. under whose ownership she was so successful on her many voyages that she was known as the "Lucky Ship," and there was no lack of men for her crew.

The end of the *Kathleen* was most dramatic. In March, 1902, she met her fate in the 12-40 Ground, which is about a thousand miles off the coast of Brazil. A large whale was seen not five hundred feet from the ship, coming head on with great speed. One of the men tried to harpoon it, but missed. The whale kept its course straight toward the *Kathleen*. As it approached, it tried to dive under the ship, but was too near and struck her forward a few feet under water. The whale came up on the other side and lay there stunned.

The captain was told that the ship was filling rapidly. He immediately lowered the remaining boats and set off in search of the others, abandoning the *Kathleen*, which soon sank.

The whaleboats were found and all started for Barbados, a thousand miles away. They soon became separated. Two boats were picked up by the *Borderer* of Glasgow, and the others reached Barbados safely after nine days.

So passed the Lucky Ship. Shall one say it was in just retribution by the leviathan that she had pursued and so often conquered?

PREFACE

The editor wishes to thank his good friend Robert B. Ferguson, for the use of his father's diaries, his notes of his father's life, and his helpful review of the manuscript.

L. D. S.

Bayside, N. Y.
Sept. 1935

CONTENTS

Chapter	Page
PREFACE	vii
BRIEF SKETCH OF ROBERT FERGUSON'S LIFE	xiii
I. FROM NEW BEDFORD TO THE WHALING GROUNDS NEAR THE AZORES	3
II. OFF GIBRALTAR AND SOUTH TO TRISTAN DA CUNHA	23
III. FROM TRISTAN DA CUNHA TO ST. HELENA	47
IV. FROM ST. HELENA TO COMORO, SEYCHELLES, JAVA, CELEBES, AND BACK TO COMORO	65
V. AT JOHANNA, COMORO ISLANDS, ZANZIBAR, AND BACK TO JOHANNA	85
VI. FROM COMORO ISLANDS AND MOZAMBIQUE TO ST. HELENA	105
VII. FROM ST. HELENA TO KABINDA, CONGO, AFRICA	127
VIII. FROM KABINDA, CONGO, AFRICA TO ST. HELENA	145
IX. AT ST. HELENA, THENCE TO MAYUMBA, AFRICA, AND BACK TO ST. HELENA	175
X. FROM ST. HELENA AS MASTER OF A SHIP TO BOSTON	199
XI. FROM BOSTON TO LIVERPOOL, ST. HELENA AND ST. PAUL DE LOANDA, AFRICA	217
XII. FROM ST. PAUL DE LOANDA TO THE SOUTH ATLANTIC WHALING GROUNDS	231
XIII. FROM THE SOUTH ATLANTIC WHALING GROUNDS TO ST. HELENA	249
XIV. AT ST. HELENA	267
XV. FROM ST. HELENA TO NEW BEDFORD	289
APPENDIX: GLOSSARY OF NAUTICAL AND WHALING TERMS	303
THE "LAY," CATCH OR WAGES	315

BRIEF SKETCH OF ROBERT FERGUSON'S LIFE

Robert Ferguson was born in Greenock, in the county of Renfrew, Scotland, on Sept. 29, 1855. His father, a master baker, thrived on hard work and required the same of his sons. His mother, whom he adored, was kind, considerate, and loving to him and the rest of her large family. There were in all eight boys and one girl, all born in Scotland except the youngest boy, who was born in America.

In his boyhood days, young Bob lived the usual life of a Highland boy, attending what schools were available and roaming the hills with his Gaelic-speaking companions.

When he was about ten years of age, the family came to the United States just after the Civil War and settled in Philadelphia. Young Bob got no more schooling but went to work in his father's bakery, where he worked until nearly seventeen years old. Through a difference of opinion which arose between him and his father, he left home. He went to New York City, where he signed up for a whaling voyage in the Arctic Ocean on a ship about ready to leave New Bedford, Mass.

Proceeding to New Bedford, he found the ship ready to leave. Thus he entered on a new career and a voyage which lasted twenty-two months in the frozen North.

For the first three or four years at sea, he kept no records, but from the years 1875 to 1885 diaries were faithfully kept. In addition to his own experiences, these records also contain some interesting tales that were told him by people whom he met on his voyages.

The following comprise the more important voyages during which Captain Ferguson kept diaries:

1. From Aug. to Oct. 1875—a sailor before the mast on the schooner *Benjamin B. Church* from New Bedford to Santiago, Cuba, and return.

2. From Oct. 1875 to Jan. 1876—second mate on the schooner *Samuel Wackrell* from New Bedford to Jamaica and back to Philadelphia.

3. From Jan. to May 1876—quartermaster on the American liner *Pennsylvania* plying between Philadelphia and Liverpool.

4. From May 1876 to Sept. 1877—harpooner on the *Abbie Bradford* from New Bedford to the Arctic (where they wintered) and return to New Bedford.

5. From Oct. 1877 to Jan. 1878—boatswain on the barque *Weser* from New Bedford to the Baltic, Gulf of Riga and return.

6. From Feb. 1878 to Mar. 1880—sailor before the mast on the New York packet ship *Mohawk Chief* from Buzzard's Bay to Liverpool. There he ran away and shipped as second mate on a small barque which was wrecked rounding the Cape of Good Hope. From there he worked his passage back to Liverpool where he shipped as mate on a brig to St. Johns, Newfoundland, returning to New Bedford.

7. From May 1880 to May 1884—harpooner on the barque *Kathleen* for a whaling voyage. On this trip he captained the barque *Daylight* from St. Helena to Boston and returned to the *Kathleen* at St. Helena.

8. From May to Sept. 1884—chief mate of the schooner *Susquehanna* from Philadelphia to Venezuela, returning to Baltimore.

9. From Sept. 1884 to Dec. 1884—chief mate of the schooner *Lancer* from Philadelphia to Turks Island and to Liverpool.

10. From Dec. 1884 to May 1885—master of the ship *Lapwing* from Birkenhead to Capetown and return.

About 1890, he gave up the sea and took a rigging job with the Pennsylvania Railroad. On Jan. 10, 1893, he married a Scotch lassie whom he met in Philadelphia. He had established himself in the grocery business in Philadelphia, from which he retired in 1921. Shortly afterwards his wife died. Much of his time after this was spent telling boys in schools, at the Y.M.C.A., Boy Scout camps, etc., of his experiences in foreign countries and at sea.

In September 1924, he returned to New Bedford for a visit, finding several old acquaintances, among them Frank Knowles who had sailed with him on the *Kathleen*. He also found his former cabin mate on the same ship, Samuel Hazzard, living at Myricks, a small town near New Bedford.

What a fine time these old friends had recounting and living over again their exciting adventures of the past!

Until recently, Captain Ferguson has been living with his son in Plainfield, New Jersey, where he died, July 3, 1935.

ROBERT B. FERGUSON

Plainfield, N. J.
Sept. 1935

FROM NEW BEDFORD TO THE WHALING GROUNDS NEAR THE AZORES

Picking the watch—Rough weather—Green hands sick—Preparing boat gear—A "stinker"—Whaleboat practice—Waterspout—Thief strung up by the thumbs—Porpoise meat—Writing letters—On shore at the Azores—Kissed by a girl—Steward taken to hospital—Brandy barrels—Porpoise oil—Gibraltar—Stove boat—Whacked by a whale—Lessons—Music—Fighting Cook—Sharks—Fouled lines—Broken oars.

CHAPTER I

THE CREW OF THE KATHLEEN
Leaving New Bedford, May 5, 1880

Captain—Samuel Howland
Chief Mate—Daniel Gifford
Second Mate—William Young
Third Mate—Robert McKenzie
Fourth Mate—John Rose
Harpooner to the 1st Mate—Frank Bishop
Harpooner to the 2nd Mate—Samuel Hazzard
Harpooner to the 3rd Mate—Robert Ferguson
Harpooner to the 4th Mate—Jose Peters
Cooper—Frank Marshado
Steward
Cook
Cabin boy

Starboard Watch	*Port Watch*
Frank Allen	Manuel King
Frank Gomez	John Donovan
James Lumbrie	Frank Knowles
John Brown	Fernand Reis
John Beebe	Frank Joyce
Manuel Sawyer	John Paul
Mark Rodesinger	Joseph King
Carlo Antonio	Otto Lubeck
Manuel Keno	Antonio Josephs
John Bravo	Julius Wing

FROM NEW BEDFORD TO THE WHALING GROUNDS NEAR THE AZORES

Tuesday, May 4, 1880
Ever since my last whaling voyage of sixteen months in the frozen Arctic almost three years back, my intentions have been to go north again at the first opportunity. I liked whaling among the cold and stormy fields of ice and, in winter when frozen in, to mingle with the natives and go hunting walrus, seal and bear on the floe ice.

While on one of several coasting trips, I heard from a friend that another whaling voyage to the north was being arranged. Losing no time, I went to New Bedford and arrived at ten o'clock this morning. I went to see Squire Butts right away. He said that on account of some mistake on the part of the owner of the ship that I would be unable to go on this voyage to the north. However, at dinner with the Squire and his daughter, he said that I could have another job if I would take it. It seems that the *Kathleen*, a small barque of three hundred tons or thereabouts, bound for a long whaling voyage of several years, needed another harpooner. Would I take it? I replied that I would.

So here I am, bound for a sperm whaling voyage for a term of four or five years, in the North Atlantic, South Atlantic and Indian Oceans with a crew that I have never seen.

As I did not have to go on board until the morning, I wrote some letters home to my folks in Philadelphia, not to all of them, for there are mother, father, one sister and seven brothers. It has been quite a while since I have been at home, because I have been on shore only five days all told between voyages since I first went to sea about eight years ago. I try the best I can to let my folks know in what ports I expect to be, but it's mighty few letters from them that come to hand. Whenever I get a chance to mail a letter home, I do. Sometimes it is months before I get the chance, like when I was wintering up in Greenland.

I am hoping that the officers on the *Kathleen* are men, not brutes, like those on the packet *Mohawk Chief*.

Wednesday, May 5, 1880

I came on board the *Kathleen* this morning. The officers seemed to know each other, but I knew no one. Not a soul did I know, officer or man. Fore and main topsails, foresail and staysail were hoisted. I helped the mate lash the anchor in board and set the fore staysail heading down the harbor. The captain called for one of the harpooners to take the wheel, not wishing to trust any of the green hands. I volunteered and the captain asked for my name. When I told him, he asked for whom I shipped to steer. I said that I was going to steer for the third mate but that I had never met him. After getting the fore and main topgallant sails on and passing Fort Phoenix and Clark's Point, the captain ordered one of the foremast hands to take the wheel. He then called Mr. McKenzie, the third mate, and told him that I was his harpooner. Mr. McKenzie came over and shook hands with me.

By this time the wind was getting strong. The ship had on all the sail that she could carry, jib, flying jib, mainsail, foresail, spanker, mizzen staysail and gaff topsail, driving out to the eastward. There were lots of coasters going north and south and one large ocean steamer bound for New York. By four o'clock after we had passed eastward of Block Island, she began to pitch and roll, making some of the green hands look squeamish. After clearing up the deck and coiling up all the rigging on the belaying pins, all hands were called to pick the watch, that is, to divide the men equally into the port and starboard watches. The chief mate, who heads the port watch, has first pick. The second mate has next pick for the captain's or starboard watch. Then it is man and man about until all are chosen. The third mate stands with the chief mate and the fourth mate with the second mate, while each harpooner stands in the watch of that officer for whom he steers.

We had a makeshift of a supper, but I was hungry and glad to get it, rough as it was. I met my buddies for the first time to talk to. We harpooners and the cooper all live in the steerage where we have ample room as the bunks are all on one side. At seven o'clock I set the watch. Unlike the merchant service where the watch is from eight until midnight, on a whaler it is from seven until eleven. No whaler rings eight bells, always eleven, three and seven. I have the first watch out tonight, a dirty, rainy night with squalls. At ten o'clock I took in gaff topsail, flying jib, topmast staysails, and clewed up the foresail and mainsail. At eleven I called the watch and had the men go up and furl the courses. The *Kathleen* seems to be a good sea boat.

Thursday, May 6, 1880

The weather is still rough and drizzly. We are heading toward the Azores with topsails reefed. Nearly all the foremast hands are seasick and feeling bad. The cook and steward do not seem to have found their bearings yet, so we have to take what we can get to eat until things get adjusted.

Friday, May 7, 1880

Today we broke out boat gear from the hold. The officers and harpooners started getting their boats ready and setting up irons, that is, setting poles to the harpoons and lances and then grinding, cleaning, and oiling them. We put boat keg, water keg, compass, boat spade, hatchet, knives, bailer and waif, each in its proper place in each boat. Then we had to coil the lines down right in their tubs and drug the nippers, that is, to fasten the nippers to the end of each whale line. We saw to it that the oars were all right, in good condition, and put them together with the mast, sail, and paddles in each boat. We were too busy getting the boats ready for whaling to send a harpooner up on the royal yard for a lookout until the afternoon.

Wednesday, May 12, 1880

Today and for the past four days, the weather has been fine with light winds and we have been cruising along with all sails set. We fastened the rings for the men on lookout on the fore and main royal masts, sent the royal yards down on deck and lashed them to the foremast out of the way. Four men were sent aloft to look for sperm whales, two foremast hands on the foremast and an officer and harpooner at the mainmast head.

Men on lookout stand there for two hours and are then relieved. They scan the ocean for a spout from sunrise until sunset.

This afternoon Mr. Rose, the fourth mate, sighted a dead whale. This is what the whalers call a "stinker" and it is well named. We lowered a boat, got the whale alongside and put on the fluke chain. The cutting gear was rigged up on the mainmast and guyed out. We started to cut in with the falls to the windlass. As it was only a small whale, we finished in two hours' time. The blubber was left on deck where it was cut into horse pieces (two-foot lengths). We cleaned the try pots and started the try works. All hands were on deck cutting up and mincing blubber. This whale smelled very strong, not like the clean, northern bowhead. The try pots began to boil fast and kept the third mate busy bailing the oil into the large copper cooler. The blubber was old and tough but it was all cut up and put into casks on deck until needed for the try pots.

Saturday, May 15, 1880

Having finished boiling oil yesterday, today we stowed it down off deck, about twenty barrels. The decks were very greasy and had to be scrubbed with lye made from the ashes under the try pots. It was too rough today to have a lookout at masthead. The men were all gathered under the spare boats to keep dry. These boats are turned upside down between the main and mizzen masts on the quarter deck and make a good shelter. The ship has been plunging and rolling heavily and the deck was flooded with water sloshing from side to side on account of the green seas slopping over the lee rail.

The green hands felt very bad and wished they were back home again. Everything looks blue to them now but they will find things harder and rougher if they live until the end of the voyage, for this is only a light gale of wind with plenty of rain. As one of the old lads was singing "Those Little Ones At Home," you could see the tears trickling down the cheeks of the new men.

When night came on, it got rougher and we got orders to hoist the boats on the upper cranes, putting extra gripes on to keep them from getting smashed or carried away. We are lying hove-to under close and fine goose-winged main topsail, keeping a lookout on top of the try works as we are in the ocean ship track.

Tuesday, May 18, 1880

The gale was over. Today there was a pleasant, light, easterly wind, blue sky and the sun shone bright. The men were feeling better and more cheerful. We broke out molasses, vinegar, beans, beef, rice, and flour from the hold, then overhauled all the stores and stowed them back.

In the afternoon, we lowered all four boats for practice, teaching the men how to handle the oars properly, set the mast and sail, and use the paddles without making any noise. Some of these green hands had never pulled an oar before. The Portuguese were all right and at home in a boat, soon learning how to pull ahead or go astern when told. All my crew were Portuguese except one, a German, who soon picked up and became right handy although he could not understand English. This made it hard for him but with the little German that I knew, we got along.

Wednesday, May 19, 1880

We had all sails set early but it got to breezing up strong and we had to clew up the mainsail and furl it. We hauled down the jib, flying jib, main and mizzen staysails, brailed in the spanker and furled both topgallant sails. In the afternoon, we had to reef and double-reef the fore and main topsails.

There was a large waterspout on the port quarter rising like a funnel. It looked just like a column of smoke rising. What a volume of water they carry up! This waterspout was about a mile off and going north. I was glad that it was no closer, for a waterspout is a dangerous customer to have for a neighbor.

We filled the scuttle butt, a large cask on deck for holding fresh water for drinking and cooking purposes. It sits aft on the quarter deck and has a spigot so the men can get all the water they want. There is no stint on a whaler. We have over two hundred barrels of fresh water in the hold, and pump it into the scuttle butt when we need it. We also keep a cask of bread (hardtack) open between decks at the after hatch so that the men can help themselves. There is no need to go hungry or thirsty at any time, but if the men waste food or water they are punished.

Sunday, May 23, 1880

Today the mate, Mr. Gifford, called me and asked me to bring the cabin boy on deck. He made the boy fast by the thumbs to a beam abaft the mainmast. The boy had a lot of money in his possession and last week told the steward that it belonged to the captain. The captain found that fifty-eight dollars in gold was gone. When questioned by the mate as to what he had done with the money, the boy gave up all but fifteen dollars. The captain said there was more. The boy denied it. Then the mate took spun yarn and tied him up by the thumbs again. The boy now said that one of the hands by the name of Paul had some and later that the cook had some. The darky cook denied it but the boy and Paul swore that he did have it. I told the captain that someone took a five-dollar gold piece out of my chest. The cabin boy said that he took it and hid it in one of the bunks. I remembered then that the cooper found it while boring a hole under my bunk yet I did not think it was mine at the time. The cooper left it on the pillow and it got lost again. The captain let the boy and Paul down after an hour but kept the negro hanging by the thumbs for another hour. The darky cook still denied knowing anything about the money so the captain told him to take his things and go forward in the forecastle. I have never thought of locking my chest on any ship that I have been on yet. Some of my shipmates have been tough old packet rats, but no thieves.

Sam Hazzard struck the first porpoise today. We got it on deck and cut up. All hands are looking forward to having some porpoise balls tomorrow. Porpoise meat is fine, chopped up with onions, not a bit fishy in taste, not oily and makes a grand change from salt beef and pork.

WHALING GROUNDS NEAR THE AZORES

Sunday, June 6, 1880

For the past two weeks we have been sailing through what is known as the Western whaling grounds and now we are only a couple of days from the Azores. Outside of seeing porpoise, grampus, a few humpbacks and finbacks, and a passing ship or two, the days have slipped by uneventfully.

Today being Sunday, all the boys were busy writing home and to their girls. One of the Portuguese, Fernand, came to me and asked me to write a letter for him to his girl. It was *some* letter. When I read it to him, he thought it was fine, but Sam Hazzard and Bishop laughed and made all manner of fun of the trash that he told me to write. I felt so sorry for him that I wrote a better one that he was very proud of. I offered to teach him to read and write. He said that if I did, he would wash and mend my clothes. He is a good lad about twenty-five years old and pulls an oar in my boat. He tells the other men who don't understand, what to do. Fernand comes from the island of Flores, one of the Azores.

THE AZORES

Tuesday, June 8, 1880

We sighted one of the Azores today and stood close in shore to a small fishing village called Fishon Legrand or Big Beans, located on the side of a high hill. The houses are mostly small, of one story, built of stone and quite strong. Some are thatched while others have red tile roofs. The walls are whitewashed and look pretty. Streams of water, clear as crystal, rush down the hillsides on both sides of the village.

As this is not a harbor, the ship could not come to anchor, for the water is deep close up to the rocks. The ship is laying off and on, that is, sailing back and forth under easy sail.

I went on shore with the captain and a picked crew of Portuguese, to trade flour for new potatoes, onions, sweets, butter, cheese, eggs, chickens, and ducks. We landed fourteen boxes of tobacco, two bolts of calico and three bolts of cotton cloth which we had to carry up the hill about half a mile to the general store. The fishing boats took all the goods we traded for out to the ship. We also got a lot of small cheeses the size of a dinner plate, oranges, apples, green figs, and a basket of fresh fish.

While we were on shore, a hard blow came up and the ship had to keep back from the rocks and the island. We had to haul our boat up out of danger and stay on shore. The captain said that we could not go out to the ship as she had to run to the leeward for shelter and told me to have Fernand find me a place to sleep.

The people on the island are clean, pleasant, and look healthy. The men as well as most of the women and girls go barefooted. As we were going up the hill, Fernand spoke to the priest. The padre could talk English fairly well and asked me to come to his house. It was a long one-story house with a tile roof and had at least six rooms. Fernand left to go back to the captain and act as his interpreter. The furniture in the house was plain and the floors were earth, packed down hard and with a few rush mats here and there. The bedroom was nice and clean but the beds were too soft, being thick feather beds. We had a good supper of fried fish, plenty of vegetables, cheese, and coarse bread. I was hungry, having had nothing to eat all day. A fine, clean old lady and her niece took care of the house and did the cooking for the padre.

The priest, a man about thirty-five years old, was very jolly and good company. He stayed home tonight and told me all about the island and the people. They burn fish-oil lamps here which smoke and smell just like those we had at home in the Highlands of Scotland.

Wednesday, June 9, 1880

Being a rainy day, the priest and I stayed indoors. We had a good talk about Portugal and the United States. The captain sent Fernand to get me and to find out how I was making out. When I told the captain I was living with the priest, he laughed heartily. But he was glad when I told him how kind the padre was to me and how well I fared.

WHALING GROUNDS NEAR THE AZORES

Sunday, June 15, 1880

A man came to the door this morning with two donkeys, one for the priest who told me that he had to go to a large village ten miles away to hold mass. He suggested that I go out for a walk and see the place. After breakfast, the girl got the priest's razor and strop so that I could shave and get cleaned up.

I took a walk over to a place about four or five hundred feet up where a small river came tumbling over the rocks into a little pond, clear as crystal, with small white pebbles on the bottom and lots of pike and trout swimming around. I sat down on a stone, looking at all the rocks and trees and watching the fish swim in the pond. I forgot all about time. At last I started down to the cottage, meeting the girl, who said, "Come to dinner." She was barefooted and had on a new calico dress. The old lady had a right good dinner and seemed glad to see me eat so hearty. Later, the priest came home and we had a long talk while the old lady knitted and the girl cleared away the dishes.

Monday, June 14, 1880

It was a beautiful morning and after breakfast when the *Kathleen* hove in sight, Fernand gathered up the men. As I was bidding good-bye to the priest at the door, I took my silk neckerchief, a very fine one about one yard square that I had bought in Liverpool, and gave it to the old lady. She put it around the young girl's neck. Then the priest said something to the girl who immediately came over and kissed me. At that, everybody laughed. One of the men who rows in my boat saw it too.

We got on board at noon and set all sail. The men on board said they had been through a fierce storm and had to run to the lee of the island of Terceira where they stayed off shore until today. We are now going to Fayal with the steward who has been taken sick.

FAYAL

Wednesday, June 16, 1880

We were hove to all night off Fayal. This morning the captain, second mate, two boat steerers, and two foremast hands went on shore to take the old steward to the hospital. About five o'clock, they got back and brought a man in his place.

There were two barques, a ship, and the brig *Abbott Lawrence* lying here. Captain Moyser of the brig sent his regards to me. He was first mate on the *Abbie Bradford* on my first voyage to Greenland. In the morning, we are going to set sail for the whaling grounds off the Straits of Gibraltar between Morocco and Portugal.

Thursday, June 17, 1880

In the past few days, we have picked up out of the sea several barrels of kerosene and a couple of barrels of French brandy. Some poor fellow must have lost his deck load. The barrels must have been in the water for a long time as they were covered with barnacles and there were lots of fish swimming around them. We also saw a few finbacks and two large turtles. We lowered a boat and got one of the turtles but the other got away. There were lots of porpoise all around as far as you could see, coming and going in all directions, jumping and skipping along, sporting and playing. After we got two, all the porpoise disappeared, not one to be seen any place. The porpoise blubber was made into oil, but kept separate, because this is a very fine oil used by watchmakers and brings five dollars a gallon.

Tuesday, June 22, 1880

Last evening, we sighted a whaler, the barque *President* of New Bedford, thirty-three months from home and with nine hundred barrels of oil. They lowered a boat and gammed with us until ten o'clock. Everybody exchanged news and had a good time. About noon today, we sighted the coast of Morocco. We could see the sand hills and two or three feluccas in close to shore. The smell of the goats on shore was quite strong as the wind was blowing from the land. Up at masthead, I could see a few scattered houses, date trees, camels, and horses.

Wednesday, June 23, 1880

This afternoon, we were near Cape Spartel about seven miles

from shore. I could see the city of Tangier from masthead with the glasses. It is quite a large city. There were many large ships and steamers going in and out through the Straits of Gibraltar. We had a good sight of the Rock of Gibraltar, the rock of fame and wonder. All around us were fishing boats and feluccas. The *Kathleen* is headed toward Trafalgar Bay where Nelson fought his last fight against Spain and France.

Friday, June 25, 1880

We awoke at five-thirty this morning to the sound of "There she blo-o-ows and there she breaches and there she blo-o-o-ows!" Mr. Rose, the fourth mate, sighted the whales from masthead and called all hands to get the boats ready to lower. Sure enough, there was a school of small sperm whales, sporting and playing like dolphins. At six o'clock came the order to back the main yard and swing the boats all clear. All four boats were lowered away. The whales being to windward, we pulled for a while with the oars and then set the sail.

Mr. Young, the second mate, got fast to a whale. Mr. McKenzie told me to dart at Mr. Young's whale. I drove the first iron into it, good and plenty. It let fly with its flukes, that is, its tail, and struck me on the seat and sent me flying from the boat thirty feet and high up in the air. It struck the bow oarsman, Fernand, a glancing blow and cut the bow of our boat clean off where I had been standing. Otto, the German, threw me a rope and hauled me to the boat, but when I got there it was full of water.

Mr. McKenzie got the whale line fouled around his ankle and was pulled down out of sight. I grasped the whale line and pulled myself down after him and cut the line. We both scrambled into the boat and upset it. The third mate's ankle was badly skinned. The whale was dead. I had killed it with my iron and now it lay quiet. This was my introduction to my first sperm whale.

Mr. Young, the second mate, took all of our crew into his boat. The first and fourth mates' boats towed the whale to the ship about two miles off. They got the whale alongside about four o'clock and started to cut in right away.

The second mate towed our broken boat back to the ship stern foremost while I sat in it to keep things together. We got to the ship about six o'clock with the broken boat where I was in the water up to my armpits. When I got to the ship, I had to be hoisted on deck. I could not stand on my feet. The men on board the ship thought that I had been killed. I fainted and Mr. Young picked me up, carried me down into the steerage, undressed me and put me in my bunk. My back was all black and blue from my seat to my neck and I could not use my legs. Mr. Young rubbed my back with something and rubbed hard until a froth gathered. I could not lie on my side or my back, only on my belly all the time.

By evening, the whale was all cut in. They hauled the broken boat on deck so the cooper could mend it. The captain came down and talked to me after things were cleaned up. He was at masthead and when he saw me go flying through the air, he thought that I had been killed. The captain spoke very nicely to me and wanted me to take a drink of brandy but I told him that I would rather have a cup of coffee. He soon had one for me and said that he was very angry with the third mate for putting me on to a fast whale when he should have put me alongside of a loose one. We lost the Pierce dart gun, oars, paddles, hatchet, sheath knife, one of the lances, harpoon, and boat compass. Our crew was green but quite cool. They did not get excited and will make a good crew. The second mate, Mr. Young, comes down every hour and relieves me of pain by rubbing my back. Poor Fernand is laid up with a sore side where the whale struck him a glancing blow.

Saturday, June 26, 1880

The cooper and the carpenter have worked steadily all day repairing the whaleboat and they made a good job of it. About dark, they finished and hoisted it on the davits. After supper, the hands started the try works and boiling oil. My back is rubbed every hour. It relieves the pain.

WHALING GROUNDS NEAR THE AZORES

Sunday, June 27, 1880

Boiling oil was finished by noon. Twenty barrels were stowed in the hold. Mr. McKenzie had to put the gear back in our boat for I was not able to help him. It kept him busy all day getting the harpoons ready and coiling down the lines. I felt a good deal better today and got on deck, resting in a chair with a pillow in it. When they started to scrub down at four o'clock, Mr. Young picked me up, carried me down below and undressed me. This big, strong man is as gentle as a woman. He rubs my back and talks to cheer me up. He is very kind to me, more so than any other of my shipmates. The last thing before going to sleep, he came down and wrapped me up in blankets and fixed me up like a mother would.

Friday, July 2, 1880

My back has been so much better that I have been able to go to masthead for a short time the past two days. While aloft today, I sighted a lone sperm whale coming toward the ship. I sung out and as it was only three-quarters of a mile away, the ship was hauled aback. Three boats were lowered and the fourth mate went in my place. The captain told me to stay at masthead and keep track of the whale and the boats. At the first rising, Mr. Young was right on its back and Sam Hazzard drove both irons into it. The whale showed fight but after thrashing the water for a while with its flukes, sounded deep. When it came up, it breached almost clear out of water and fell over. Then it came at the boat head on and full of fight. When it neared the boat, it rolled over on its back with its long under jaw showing and snapped its large teeth at the boat. The boys had to be quick with the oars and back away. I had a fine view of the fight from masthead, as the whale was only a quarter of a mile away. Finally Mr. Gifford, the first mate, with his loose boat got alongside and fired a bomb into it. The whale gave a loud snort like the sudden exhaust of a locomotive, then shot forward and rolled over on its side close to the ship. I thought the whale would get Mr. Young's boat but the Portuguese boys were there every time with the oars. The whale was cut in by dark. We finished boiling the next morning and stowed down thirty-eight barrels.

Sunday, July 4, 1880

Today was Sunday and our national holiday. In the afternoon, I gave three of the Portuguese and Otto, the German, lessons in reading and writing. Otto is quite a scholar and helps me with things above my reach, explaining them so that I can understand. I enjoy talking to him for he knows a lot of history and many things that I cannot grasp very well. I told him how limited my education was and how I had to go to work when I was only nine years old and should have been going to school. We gammed the whaling barque *President*. The Portuguese had a good time playing the accordion and the guitar and dancing and singing. Otto asked me to get the loan of the accordion and said that he would play for us. So Fernand got it and Otto played "The Star-Spangled Banner." It sounded so good that all the men gathered round and the officers came out of the cabin to listen. I told Otto to play "The Watch On The Rhine." My, but that was fine! After supper, all of us boatsteerers sat around the vise bench, talking about where we were a year ago. Sam Hazzard was on the barque *Cleon* in the Arctic Ocean. Frank Bishop was on the barque *Desdemona* in the South Atlantic Ocean. Jose Peters was in the Pacific Ocean off the coasts of Peru and Chile, on the barque *Falcon*. As for me, I was in the North Atlantic. Now here we are, gathered in from the four quarters of the globe on the barque *Kathleen*.

Thursday, July 8, 1880

We are still cruising the whaling grounds off the Straits of Gibraltar. At noon today, the cook had some trouble with some of the foremast hands. Three of them jumped on him. The cook is a southern darky and a tough, thick-set little fellow. He blackened the eyes of one of them and I believe would have mauled the three of them. The Portuguese cannot fight even in a rough-and-tumble. Mr. Gifford sent the three to masthead until dark for punishment.

Mr. Young still rubs my back three or four times a day, but it is getting better. I feel as if I ought to stand watch but the captain says I should wait until my back is stronger on account of being likely to wrench it. He thinks I am lucky not to have been killed.

Sunday, July 11, 1880

We have been having a spell of wonderfully good weather. There are always plenty of large sharks hanging around the ship. We hauled one in on deck, for we wanted its skin for sandpaper. It makes the finest sandpaper, especially on ivory or hard wood. I had a group of men around me all afternoon learning their lessons. Later at masthead, I could see the people walking about the small town on shore and a lot of small dhows going up and down the coast. I could see a few camels going along rocking like a boat, and men on horseback with the trappings glittering in the sun. Their fine horses pranced as if they were too proud to walk. The men were all dressed in white, with red sashes around their waists and large white turbans on their heads. I could also see the date palms waving in the breeze, real plain.

Tuesday, July 13, 1880

Weather fine and clear with a land breeze, which you could tell by the smell of the goats. About ten o'clock, I saw two whales near shore about a quarter of a mile off the sand dunes but thought at first that they were killer whales and did not pay much attention to them. Finally Mr. Young put the glass on them and his voice came ringing out, "There she blo-o-ows!" again and again. They were sperm whales and only a half a mile from the ship, sporting and playing in shoal water. We hauled aback the main yard and lowered three boats very quietly, making no noise. All hands were barefooted and had only their undershirts and dungarees on. We up sail and paddled to where they were sporting. Mr. Young and Mr. Gifford both struck at the same time. I had to stay at masthead looking out for the boats and keeping track of the whales. Sam Hazzard set his whale to spouting blood and kicking around some but there was not much fight in it and as it could not go down very deep, Mr. Young soon killed it with a lance.

But Mr. Gifford's whale ran off and up the coast for five miles. It fought and swung its tail in all directions. It continually snapped at the boat with its long lower jaw. Then it turned offshore and sounded in deep water, finally coming up with a breach almost out of the water. It bit and kicked and kept the men busy with the oars to stay clear. The chief mate lanced it good and it went into the death flurry. Mr. Rose, who had been following, came and helped Mr. Gifford tow it to the ship.

It kept seven men busy killing sharks around the two carcasses. We started to cut in right away. The deck was all littered up with blubber and the two heads. I stayed up on deck until midnight keeping a lookout or at the wheel. We don't often see sperm whales in shoal water, mostly in deep water close to the rocks. We got forty-eight barrels of oil from the two whales.

Saturday, July 17, 1880

While we were gamming the barque *President*, a school of whales was sighted and the ships hauled aback. Three boats were lowered from the *President* and all four from the *Kathleen*, all chasing into the middle of the school. The *President* got two and Mr. Young got one. Later all of the boats were fast to whales, but the irons drew out of some and others got twisted fast. The lines got fouled so badly that some had to be cut to keep the boats from going down. It would have been a fine day's work if we had got all the whales that we struck, at least seven or eight for both ships. All three whales were taken alongside the *President*. The oil will be divided between us.

Monday, July 19, 1880

It's fine weather with a good breeze from the north. The *President* is close by. Before breakfast, we raised whales to the windward, lowered the boats and pulled all day after them, but could not get near enough to strike. Two or three times I stood up ready to dart but had no chance. It was the same with the

WHALING GROUNDS NEAR THE AZORES

other boats. We got back to the ship at six o'clock at night, tired and hungry, as each of us had nothing all day but a drink of water and a hard-tack. Mr. Young came down and rubbed my back and it felt mighty good after pulling an oar all day. He told me to stay in my bunk all night and that the captain had said so.

Tuesday, July 20, 1880

We had just finished breakfast when all hands were called to lower the boats. We pulled up to windward, set the sails, and Mr. Young, first of all, got fast to a whale. The third mate put me on to one and I drove both irons into it clear up to the hitches and set it spouting blood. The first mate was close by and he sung out, "Good boy, Bob, that's the way to get revenge." I laid the boat close up to the whale with the steering oar and Mr. McKenzie killed it right away. The chief mate killed his whale and Mr. Young helped the fourth mate kill his. We towed the four whales to the ship one at a time and got them all fluked alongside by ten in the morning. That was quick work. We started to cut in right away with a smooth sea, clear weather, and no sharks around. Sam Hazzard had his arm hurt while cutting in. When the first mate got fast, his bow oarsman, Manuel King, got a bad bump on the head when the whale struck an oar and broke it. The rowlock broke and one of the pieces of the broken oar struck him and came near killing him.

A light squall for fifteen minutes when we were cutting in split the flying jib from sheet to clew, but we hoisted a new one. A passenger steamer came down close to us to let the passengers get a good look. Her rail was level with our maintop.

Wednesday, July 28, 1880

Last Sunday, Mr. Young's boat and ours got two more whales. On top of this, I got a small calf in which I stuck a waif, but the little fellow sank. We have changed our course now and are heading to the southward, running along close to the shore. We sighted a lone whale on the port bow and lowered all four boats. At the first rising of the whale, Mr. Gifford got fast and had his boat stove in. He had to get in our boat and sent me to the ship with his. The whale was finally killed after it had milled around for a long time, towing all three boats after it. It seemed to be nothing for the whale to tow those three boats which were hard put to keep clear of its head and tail. After the mate shot a bomb into it and it spouted thick clots of blood, he was able to get it alongside.

OFF GIBRALTAR AND SOUTH TO TRISTAN DA CUNHA

Tangled lines—Swarms of sharks—Killer whales—"Bloody Dan"—Teaching school—Checkers—Darky songs—On shore at Fayal—Transferring oil—A new cook—Hands run away—New hands—Almost wrecked—Cape Verde Islands—Man killed from aloft—Burial at sea—Portuguese pie—King Neptune—Cabin boy spanked—Stove boat—Gales—Blackfish—Albatross.

CHAPTER II

OFF GIBRALTAR AND SOUTH TO TRISTAN DA CUNHA

Saturday, July 31, 1880
We are still coasting south off the African shore and today passed some of the Canary Islands. The last lot of whales that we got gave us one hundred and three barrels of oil. The bow boat that was stove in is now in good order.

Sunday, August 8, 1880
We set all sail at daylight, down tacks and sheets and nothing in sight. It's fine and clear with a light wind from the land. The men are lying in the sunshine all over the deck, on the fore hatch and forward of the windlass, sleeping or telling stories, reading or playing cards, checkers, or dominoes. I gave my Portuguese scholars a good lesson all afternoon and had some English and German with Otto. It's wonderful how quickly Otto has picked up the English. He speaks it fairly good now and understands a great deal. I can hold a conversation with him and he speaks correctly. He is very fond of me. The captain asked how I was getting along with my scholars. I told him what a good scholar Otto was and how much he helps me in my figuring and geometry, that he is a good man in the boat, pulls a strong oar, learns quickly, and is a fine carpenter.

Monday, August 9, 1880
We were breaking beef, pork, beans, flour, and pickled onions out of the after hold, when John at masthead sung out with a thrilling voice, "There she waters, there she white-waters and there she breaches!" and again "There she blows, and there she blows. Sperm whales on the lee quarter." It was nearly five o'clock in the afternoon. All four boats were lowered away and we chased a school of sperm whales with our sails and paddles so quietly that we got right in the middle of them.

By six o'clock each of the four boats was fast to a whale. Then all of the lines got fouled. The lines of the second, third, and fourth mates' boats got tangled around the chief mate's whale. All we could do was to cut the lines to keep the boats from going down. Since the whales and boats were all grouped together and could not get out of the way, the whales would have soon made matchwood out of the boats if they had ever struck them. One whale was mad and on its back running after the boats snapping its jaws. All of the whales were swinging flukes right and left. Mr. Gifford shot a bomb into his whale and stopped its career of snapping. We got fast again but the iron drew out. Then the third mate put me on again and I got both irons in and set it spouting blood.

We got all four whales alongside the ship, a half hour after dark. The whale line was wrapped around and around one whale eight or ten times. The mate's boat had the gunwale smashed by the fourth mate's boat on account of being unable to slack the line, which was jammed around the loggerhead. The last whale got a crack at him and split one of the planks in his boat.

When the boats were all hoisted on the cranes and all fast, we had supper at half-past ten. The watch was set until daylight and those who were on watch had to get out the cutting gear, falls and guys to the windlass, and the cutting stage over the side with the guys hauled out. They had to get junk casks out of the forehold and set the bug-light over the cutting stage so as to see and kill the sharks. I must have killed at least forty sharks in the two hours that I was out on the stage. The bug-light is a hoop-iron basket and burns whale scraps. It gives a very good light. We hang it over the side above the whales so that we can see the sharks and cut them in half with a cutting spade, which has a handle from ten to fifteen feet long and an edge as sharp as a razor. When standing on the stage, you can easily reach them. If you cut a shark on the nose, it kills him instantly. There are always at least two men killing sharks when cutting-in has to be carried on through the night, but during the day the men working on the whale take care of the sharks.

TO TRISTAN DA CUNHA

Tuesday, August 10, 1880

Called all hands at four-thirty to man the windlass and got busy cutting up the whales. One man is turning the grindstone for the cooper who is grinding and sharpening the cutting spades. The captain and the first two mates are on the cutting stage scarfing the blubber and peeling the blankets from the carcass as the men under the third mate heave the windlass. The fourth mate and the boatsteerers are in the waist taking care of the blubber. It comes to them in large blanket pieces in which the fourth mate cuts a hole in the blubber with a large knife like a sword, a hole large enough so that the large, heavy hoisting hook can be put in. We hoisted the heads on deck and separated the junk from the case and got all cut in by supper time. When we let the carcasses go, there were swarms of sharks around, hundreds of them.

Wednesday, August 11, 1880

After breakfast the harpooners started straightening out their boats and putting things in shape after yesterday's mess. The lines had to be overhauled. The gear was all foul. Lances were bent and so were the harpoons. New poles had to be stepped in the irons. The lines had to be coiled down in the tubs. We had to get out more casks from the hold to cool the oil. As the casks shrink with hot oil, the cooper has to drive up the hoops on the casks several times to keep them from leaking. Sometimes the oil is too hot to stow away, so we have to let it cool down. To make room for the new oil, sometimes we have to pump out the fresh water from some of the casks in the hold and run down the oil to these empty casks in a hose.

Saturday, August 14, 1880

For the past ten days we have been working to the north. Today, we are off the Straits of Gibraltar again. We sighted a large school of algerine or killer whales, extending as far as you could see in all directions. They were sporting and playing like porpoises. The captain lanced one from the bow but the killer took lance and warp and got away with it. The mate lowered

a boat and got the one that the captain lanced and it ran its nose through his boat at the stern. We hoisted the killer whale on deck, opened it and found a large bone in its stomach. Killer whales go in schools like wolves and often attack a large whale when they find it alone. They hang on to it and worry the whale to death. The killers run from ten to twenty feet in length. We boiled out the killer in an hour and got one barrel of oil, which is classed as blackfish oil.

Sunday, August 15, 1880

I gave John Brown and Frank Gomez their usual two-hour lesson. At dog watch, Otto and I had a lesson together. He read two chapters in my Bible and did very well. He has shown me how to do some figuring in the navigation lessons that I am working out and studying. I take a sight now and again and the captain encourages me.

Tuesday, August 17, 1880

We set all sail at daylight with a very light wind. We are off the coast of Portugal not far from Lisbon. The men are making mats for chafing gear on the backstays in wake of the yards, some thrum-mats and some sword-mats. Tonight it is such a fine, clear, moonlight night with not a cloud in the sky that every one is up on deck. There is something fascinating about a clear moon out on smooth water and no noise but the murmur of the men's voices. On a night like this, something steals over one that brings thoughts and visions from afar off. Some think of their friends or of a girl that was left behind. My thoughts went wandering down to the West Indies and a moonlit night in an orange grove. Yet I have no longing to be anywhere but where I am. I am quite contented and have good companions. I hope to visit some of the places that I have often wanted to see. I have no complaint. I have a good captain who holds the rough officers within bounds. The other officers are all civil. Mr. Gifford has a hard name and is known throughout the whaling fleet as "Bloody Dan," but I can find no fault with him. I have met some hard, cruel lads on some of the packet ships but, thanks to my early training, could hold my own with many of the bullies of the Western ocean. They always treated me all right for I always jumped to my duty and never shrank from any tasks set to me. Sometimes the food was poor, fit only for hogs, or spoiled by poor cooks. On the *Kathleen* the food is good and plenty, and we live better in the steerage than in the cabin of many merchant ships.

Sunday, August 29, 1880

We are now on good whaling grounds to the eastward of the Azores and westward of the Canary Islands. There are four men at masthead scanning the horizon for whales. For the man who raises a whale, there is a bounty of five dollars, provided it is caught. I was up aloft until noon. Then the wind died out, leaving a calm and glassy sea.

I held school with Mr. Young laughing at me, telling the third mate that he ought to join the class and help the professor. Mr. McKenzie said that I had better go to school myself instead of teaching.

Some of the Portuguese were mending their clothes and others were playing checkers. One black Portuguese beat all the foremast hands at the game, and he certainly knew how to play. I found that Otto also knew how to play, so I arranged a game between him and the big black Portuguese. I figured that Otto would know how to play pretty good. Mr. Gifford said that Otto could not beat that bravo. All hands, except the man at the wheel and those on lookout, stood around watching every move, even the captain. You could have heard a pin drop. The game lasted two hours and Otto won. All the Portuguese stood around with mouths open. The captain said he would like to try a game with Otto some day. The cooper told me that our captain, on the last voyage, beat all the other whaling captains in St. Helena and won fifty dollars.

Tuesday, August 31, 1880

The day started in fine, warm, and pleasant but after dinner the wind became stronger and stronger and squally. We had to take in both topgallant sails, flying jib, main and mizzen topmast staysails, and gaff topsail. We clewed fore- and mainsails, brailed in the spanker, double-reefed the fore and main topsails, hauled down the jib, furled all the light sails and she lay snug. A few stormy petrels were darting to and fro across the stern. They are little black-and-white birds, a little larger than a swallow, very pretty, quick as a flash, and follow the ship nearly all the time.

Thursday, September 2, 1880

After supper the boys, feeling good, started dancing to the accordion in the hands of the black cook. Oh boy, how he can make it play and make it talk! He played all Southern tunes like "Old Black Joe," "Swanee River," "Dixie," and "The Yellow Girl From Texas." Some of the boys sang with him and how their voices did ring across the water. I asked the cook if he could play any Scotch tunes, and right away he played "Annie Laurie," "Loch Lomond," and "Welcome To Prince Charlie." He told me there was an old Scotchman down in the State of Georgia who taught him how to play. He lived with the old Scotchman's family and did the chores. The old man, who was very strict, would get his fiddle down and play every night except Sunday. One day he brought the cook an accordion and taught him how to play it. Those were happy days, but when the old man died the cook lost his home and became a tramp. The captain asked me who was singing and was much surprised and pleased at a darky who could sing Scotch songs with a Scotch dialect.

Tuesday, September 7, 1880

Yesterday we picked up a barrel of kerosene all covered with barnacles. Today, masthead reported a long piece of timber with lots of fish hanging around it. We hauled aback, lowered a boat, and caught a lot of fish, mostly sea bass. The timber was of teak, eighteen inches square, sixteen feet long and in good condition.

We sighted the top of the mountain on the island of Pico, sixty miles away, over the top of the clouds. The mountain is very high and steep and deep water lies close in shore. Our

TO TRISTAN DA CUNHA

cooper's people all live here. They make their living by fishing and making wine. Grapes grow all over the mountain. The clusters are very large, weighing from three to four pounds, and the wine is said to be equal to that of Madeira.

FAYAL

Thursday, September 9, 1880

Yesterday we passed close to the islands of Terceira and St. George, and at sunset were under the peak of Pico. Today we anchored at the port of Fayal. There are no wharves here, just anchorage. It is not much of a harbor, just a roadstead, but it will be a good harbor when they get the breakwater finished, for which they have plenty of rocks quite handy.

The captain and the cooper went on shore and sent the boat back with the ship's letters. Of the four harpooners, only one got any letters and that was Sam Hazzard, who got four. The captain sent aboard new potatoes, onions, vegetables, fish, fresh meat, butter, cheese, and a box of fresh eggs.

I sighted the barque *Osprey* of New Bedford, the barque *Tropic Bird*, and the *Romance*, a fine large barque from Windsor, Nova Scotia. The latter was carrying passengers bound for Australia. From what I see of this place, it is beautiful, high lands and steep rocks. It is quite a busy little town with plenty of shipping. Fine, sweet oranges, apples, figs, and a crate of chickens all came off after supper. They tell me that chickens are very cheap and that the place is healthy and where one can live easy. There is no frost, no snow, no excessive heat and always a cool breeze, but at this time of year they get severe squalls.

Monday, September 13, 1880

We were busy today breaking casks of oil out of the hold. We sent 416 barrels on shore to go home on the barque *Veronica*. The after hold had to be overhauled. The cooper was busy setting up shooks. We stowed down the empty casks for fresh water and stowed away fresh eggs, vegetables, and fruit in the fore hold. The barques *Mattapoisett, Andrew Hicks, Daniel*

Webster, Robert Morris, Morning Star, Bayliss, and the whaling ship *Europa* are all here.

Wednesday, September 15, 1880

As the port watch was going on shore today, I thought I would clean up and give the place a look-over and go to a dance in the evening. I spoke to Frank Gomez about it but he found out that there was no dance hall nor any place of amusement here. Frank wanted me to go along and get acquainted with some fine Portuguese girls but I told him that I was going aboard at nine o'clock.

This is not a big town, just a busy little place, high up on the rocks with trees and flowers in abundance. The houses are good and strong, all white, and with roses on lattice work at the windows and doors. There are small but pretty gardens in front of them. As I was going up the hill, a man with a herd of goats came along singing, "Latè, Latè." That means milk. The man would stop at your door and milk the goats directly into your pitcher or can and then go on to the next customer. After watching him for a while, I went on and came to a large, open market place. There was a grand display of vegetables, red and white cabbages, turnips, carrots, tomatoes, beets, large onions and fine, large, red and white sweet potatoes. Everything was arranged in a tasteful way and the stalls were attended by neat, clean young women who were very healthy looking. Further along was fruit: oranges, apples of different kinds, green figs, lemons, peaches, melons, pears, guavas, and grapes both white and red. I saw one bunch of grapes on the scales that weighed three and one-half pounds. The grapes were sweet and fine. The plums were delicious. A little further along, I came to the fish market and saw large fresh rock cod, haddock, flounders, Spanish mackerel, sunfish, sea bass, and a lot of other fish, not to speak of shrimp and a few crabs. There are very few shellfish, like oysters or clams, as there is no sandy shore or beach around these islands.

When I came back from the outskirts of town, I met the captain all alone. He asked if I had had any dinner. I replied that

TO TRISTAN DA CUNHA

I had not as I had not brought any money with me and that I was going on board for supper. So he said to come along with him to the hotel. After dinner, we sat talking for an hour about Greenland. He asked me how come that I came with him on such short notice and wanted to know if I was married or had a girl. I told him not as yet.

Friday, September 17, 1880

Yesterday and today I was busy helping the cooper set up shooks. Some of the men were busy painting the ship dark green, striped with a white ribbon. I touched up the name on the stern and the mate said I did a good job of it.

There were seven men who did not come back to the ship this morning. They ran away. They were Frank Bishop, a harpooner; John Brown, the darky cook; and five foremast hands.

Sunday, September 19, 1880

A fine pleasant day and the port watch went on shore until Monday morning. I told Sam Hazzard to go in my place. Poor Sam had no money so I loaned him a sovereign. As Mr. Gifford, the mate, planned to go on board one of the other ships for dinner and the steward had gone ashore with the port watch, that left only me and seven men forward. I told Mr. Gifford that I wanted to stay on board and if he would let me go down in the pantry, I would get up a dinner for the men. Laughing, he told me to go ahead and that I would probably have to eat it all myself. I got Frank Knowles to go in the galley to show me how to work the range and keep a good fire. Frank put on a pot full of potatoes and when they were done, peeled them. I added a lump of butter, two eggs and a half pint of milk, salt and pepper and beat the potatoes to a cream. I made two pans of biscuits and they were light as a feather. Put a roast of beef in the oven and a small fresh ham scored and stuck full of cloves. I made three pies, one custard, one apple, and one peach, making the crust with butter and washing it with milk and egg. Made some good coffee, hunted up a clean tablecloth and polished the water tumblers, silver knives, and forks. I forgot to say that I put some sweet potatoes in with the roast and told Frank to watch and not let them get too brown. We also had fried onions, and there was celery and fruit on the table.

The mate had not yet gone out for dinner but was in the after cabin reading, but when he smelled the cooking, he said, "Bob, I am hungry and will stay here for dinner. Set the table for four and have dinner in the cabin. When will it be ready?" I replied whenever he wished it. He said that he would be back in ten minutes. While he was gone to one of the other ships, our captain came aboard. He was surprised to see the table set so neat and looking so clean. Just then Mr. Gifford returned with two mates from the other ship. I set the dinner down to the four officers and served the same dinner to the men forward. The captain said the dinner was as good a meal as he had ever sat down to. The other mates said that they were glad that they had come on board and that I surely knew how to cook.

Monday, September 20, 1880

The cooper finished setting up shooks, got beckets on sixteen casks and towed them ashore to raft off the fresh water for drinking purposes. The steward has not shown up yet. Otto ran away last night. The large box that the mincing machine usually stands in, a box eight feet long by four and one-half feet wide and one foot deep, built of two-inch plank, had been lowered over the side. It was as much as two men could lift so he must have had someone to help him. He took a paddle from one of the boats and got on shore with all his clothes. Later we found the box on the shore and towed it back to the ship. They tell me that Mr. Young, the second mate, is on shore drunk and refuses to come aboard. The fourth mate cannot be found anywhere.

With the other men gone, Mr. Gifford has to look out for all the stowage and all the work on deck. He is very grouchy about it. You can't blame him, for Mr. McKenzie, the third mate, got his discharge and that leaves him with everything to look after.

Tuesday, September 21, 1880

The *Fannie Burns*, a small whaling schooner of New Bedford, came to anchor this morning. She has 490 barrels of oil on board, all she can carry and everything filled up, ready to go home. Mr. McKenzie, the third mate who was discharged, came aboard and took his things ashore and never as much as said good-bye to me, his harpooner, for what reason I do not know. Mr. Young came on board late last night. When he was going on shore again tonight, he asked me to go along with him and said he would show me a good time with some good-looking Portuguese girls. I thanked him and said if it was a dance that I would like to go, but not if it was only a house party. So away he went, a mighty good whale man, for a riotous time, drinking and carousing all night. He is a good, kind-hearted man, although rough.

Thursday, September 23, 1880

The second mate is still on shore, drunk as a fiddler. Another of the crew, a boy of seventeen, came on board today as full as a tick. I suppose some of the older ones thought it was fun to get him drunk. They sweeten the ardent liquor with honey. It tastes good and they don't realize they are drunk until they begin to stagger.

The *Veronica*, a fine large barque, clipper built, of New Bedford, came in to anchor at noon. Tonight I went ashore to see Mr. Pollard who was on the *Abbie Bradford* when I was up north in Greenland. Jim Smith, our old cooper, and Bill Snell, harpooner on the *Abbott Lawrence*, were there too. All three were glad to see me. We talked over old times.

Friday, September 24, 1880

We hoisted 140 barrels of fresh water on deck today and stowed them down in the hold. We also got sixty bushels of sweet potatoes, twenty-five bushels of fine large onions, and 150 bushels of white potatoes. From the *Veronica* we got three tubs of butter from New Bedford, a barrel of hams, and some letters and newspapers. The mate asked me if I would act as steward until such time as the captain could get one. He told me that the captain was trying hard to get someone, but no luck yet. I agreed to do it until they could get a regular steward. So I make two pans of biscuits for the men every other day and every night for supper in the cabin. They have ham and eggs for breakfast and flapjacks every other morning. I make Dutch toast (bread dipped in egg batter and fried in butter) twice a week, gingerbread three times a week, and give them fresh fish on Friday.

Sunday, September 26, 1880

All the officers were on shore except Mr. Gifford, Sam Hazzard, and myself. I gave them a real good breakfast. The mate asked me if I had ever been steward before. I replied that I had not, but had been quartermaster on the steamship *Pennsylvania* of the American Line. Mr. Gifford wanted me to get up a real good dinner as the captain was going to bring some other skippers on board. I made a roast of beef with sweet potatoes baked in the pan, mashed white potatoes, parsnips, mashed turnips, two apple pies, and a custard pudding with dates in it. Put oranges, pears, and apples in a dish and grapes in a glass bowl for the center of the table. I also sliced some peaches to have with cream. I hunted around for napkins and at last found a dozen that had never been used. At twelve o'clock the captain came aboard with two other captains, Mr. Young, and a chum of Mr. Gifford's, and asked me if I had anything good for dinner. I said "It will be ready at one o'clock as that was when I understood you were coming on board." After dinner was over, the captain of the barque said that was not a dinner, it was a banquet. The captain of the *Europa* asked Captain Howland where he got a hold of such a steward. Our captain replied that I was one of the harpooners.

Tuesday, September 28, 1880

It was raining this morning when all hands were called. We took the gaskets off the topsails and hoisted them, loosened the topgallant sails, hoisted the fore staysail and jib, and got all the running rigging down on deck and all clear. We hove the anchor, got it to the cathead, kept off before the wind and set fore and main topgallant sails. We are going to keep tacking back and forth in front of the harbor until we get some men to replace those that ran away. Tonight we got a cabin boy and four men. It is very rough tonight and all the men are sick.

Wednesday, September 29, 1880

Several new men came on board the ship last night and all of them are seasick this morning. A fellow by the name of Charles Reed, formerly second mate of the barque *Lagoda*, came to take the place of the third mate who was discharged. An Englishman by the name of Tommy Wilson is the new man who is going to be the mate's harpooner. The captain has been unable to get a steward yet, so that falls to my lot for a while.

Sunday, October 3, 1880

We have been sailing back and forth away from the harbor for the past few days, hoping that the gale would stop. It has been severe and the green men have all been sick. This morning there was a heavy fog. The weather cleared up all of a sudden and there on our lee beam were the rocks towering overhead. Amid the roar and the dash of the breakers, it looked like it was going to be the last of the *Kathleen* with all hands.

I sung out, "All hands jump. Down fore tack and main tack. Let go clew lines, buntlines, leech lines. Come lively, boys!" The second and third mates came tumbling up on deck, hollering, "Haul off fore and main sheet!" Mr. Gifford himself grabbed the wheel. One could feel the ship jump from under. She heeled over to port with the lee rail level with the water. I thought the masts would strike the rocks, if not the masts then the keel, but it is deep close to shore. We shot past the Point of Delgada all clear and into the harbor mouth where we let go tacks and sheets, clewed up and hauled aback. It was a close call for us.

The captain came on board with more fresh vegetables, a coop of chickens, and a box of eggs, but no steward. He immediately gave orders to set sail and head toward the south. Captain Howland was hungry as he had no dinner. I brought him some supper and he told me that he could not get a steward here. He was going to try to get one at the Cape Verde Islands where we are going next. Both the captain and the mate said that everyone was pleased with my cooking and thanked me for doing this work until they could get someone else.

Sunday, October 10, 1880

The men have been busy getting the boats ready for action, but so far we have not sighted a thing except a school of grampus. The new hands are mostly young fishermen from eighteen to twenty years old, bright and healthy. Every one of them is good in a boat and pulls a good oar. Today we sighted the island of St. Vincent, a very high island, on which you could see, high up on the rocks, a man's face which looked just like that of Washington. The British have a coaling station here. The ground is not fertile, and as there has been no rain for four years it is hard for the local Portuguese to eke out a living.

FOGO, CAPE-VERDE ISLANDS

Tuesday, October 12, 1880.

Today we ran in close to the islands of Fogo and Brava of the Cape Verde group. We hauled aback the main yard and lowered the starboard boat with the captain, the fourth mate, and some of the men whose homes are here. The captain and the American consul came on board with some men, a boy, a cook, and a steward, and a new fourth mate. The former fourth mate refused to come back on board and the captain let him go. Later the captain went ashore and sent off some watermelons, oranges, ducks, and two little black pigs. Mr. Young, the second mate, has been sick and I feel sorry for him as he has been very good to me.

Sunday, October 17, 1880

We set sail from Fogo on Friday. I was very glad that the captain got a steward so that I could go back to my regular work and look out for my boat. I told the new steward that if he needed help at any time to let me know and I would help him. Tonight, up came a sudden squall. Mr. Reed told me to get the flying jib off of her and then clew up the fore topgallant sail. Tom Wilson attended to the main topgallant sail while Mr. Reed hauled down the gaff topsail. Mr. Reed told a couple of men to jump up and furl the main topgallant sail. Just as I saw that the jib and the fore topgallant sail were all snug and as I was coming aft, I heard an awful thud. Looking around in the dark, I found a man huddled up in the scuppers, having fallen from the topgallant yard. I called a couple of men to lift him on to the after hatch and told Mr. Reed what had happened. Mr. Gifford and I worked on him for two hours but there was not a spark of life in him, although no bruise nor mark was to be seen, nor was there a broken bone in his body. This man had gone aloft with one of the old hands, but Mr. Reed did not realize that this poor fellow was one of the new hands or he would not have let him go aloft at night. We carried the body to the after house by the wheel and covered him up until morning.

Monday, October 18, 1880

This morning, we took the man who fell from aloft and sewed him up in an old piece of canvas, with iron at his feet to sink him. We laid the body on the gangplank, shoved it out, hauled aback the main yard and when the captain said, "All ready," Mr. Gifford tilted the board and let him slide over the side to a sailor's grave, without a word or a prayer or a funeral service. That was the end of a promising young man in the prime of his youth, down to the depths of the ocean where no gravestones mark the place. When we harpooners gathered in the steerage at dog watch, we talked of the way that the young fellow was dumped over the side in silence. To change the appearance of things, Tommy Wilson with his grand voice started to sing, "Oh, Bury Me Not in the Deep, Deep Sea." We wondered how this poor boy's mother would feel when she gets the news a year from now or maybe longer. It is hard to tell if she will hear at all.

Thursday, Oct. 21, 1880

Yesterday we sighted the barque *John Carver* whaling, so we lowered four boats. I struck a whale. Mr. Young got fast to another and Mr. Gifford's harpooner missed one. After killing our whale, we put a waif on it and tried to get another but could not get near enough. The ship picked up the whales and we started to cut in, finishing by sundown. We had a good day's work.

Today we had wet weather, so after starting the try works, we set a smoke sail over it. Everybody was busy cutting up blubber and putting it down in the blubber room. We roasted some large sweet potatoes in the hot oil. Mr. Reed went down in the cabin and brought up a pie made by the new steward, but the crust was as hard as a piece of hard-tack and had no taste to it. I told Mr. Reed that it was a Portuguese pie. He said it was only fit for pigs, and *they* got it.

Tuesday, October 26, 1880

We broke out two casks of clothing, containing blankets, shoes, boots, pants, shirts, underwear, caps, stockings, coats, oilskins, and sou'westers. These were distributed among the new men, some of whom never had shoes on their feet before. Some of the older men played tricks on them when they got the sea boots on, like telling them that the mate wanted to see them right away. The poor fellows would try to hurry aft holding on to the rail, tumbling when the ship gave the least roll, with all hands laughing at them.

Wednesday, October 27, 1880

This day came in with a heavy downpour of rain. All hands got busy catching rain water to wash their clothing, wearing nothing but a pair of short pants in the warm rain. There are lots of sharks, big ones, all around the ship. We lanced several of them and cut some nearly in half with deck spades. Someone took a bright tin can tied to a string and dropped it overboard. As fast as a shark would make a dive for the can, one of the men would cut the shark with a long-handled spade. These sharks are the greediest things that I have ever seen. Some of them are very big and look fierce with their green eyes. They are so bold that you cannot frighten them off in any way.

Sunday, October 31, 1880

We have been steadily making to the southward, and today while I was on lookout at masthead, Mr. Gifford reported that we were five miles south of the Equator. Just then a mysterious voice came from forward crying, "Ahoy, ahoy, what ship is that?" All of a sudden, King Neptune, with his long hair and beard of manila, came over the bow. He was all shining and tattooed or rather painted with bright red and green paint all over his body. All he had on was a pair of shorts. He had three men with him and they called all hands. Eight of the deck crew were called on to be shaved. The large deck tub was full of water. There was a bucket of grease and a pot of tar too. The eight men were lathered and shaved with a hoop-iron razor. Some of the men were spanked and the little cabin boy had to run through the hurdles to get spanked. But I had fixed a piece of canvas with several tacks in it and another piece of canvas stitched to it to keep the tacks from coming out and told the cabin boy to fasten it to the seat of his pants and not to tell anyone. The first fellow that slapped the boy yelled and a dozen of the men had their hands punctured with the tacks. The boy got the best of them at their own game. When I got down from aloft, the fun was all over, so I gave Frank and John their lessons and read my Bible for an hour at dog watch.

Wednesday, November 3, 1880

The ship is heading SSW with all sails set. The weather is warm but very pleasant. We are painting the whaleboats, the bottoms green, bodies white, and gunwales blue. Our boat has a spritsail with a gaff topsail attached to it and it works pretty good. In a light wind, the topmast hoists and stretches the sail.

You have to know how to handle it when going on to a whale with the sail up, so as to get it down quick and rolled up around the mast and passed aft out of the way. When you strike a whale, there must be no bungling and everything has to be kept clear of the whale line, for it goes a-humming out of the chocks forward.

As each boat was painted and put on its crane, Sam Hazzard put all the boat gear in place. After painting Mr. Reed's boat, the captain asked me to paint his, on account of doing such a good job on the third mate's boat.

We sighted a large full-rigged ship today with all sails set, heading on the port tack. She had an awful spread of canvas, three skysails and stunsails on the main. Deeply laden, bowling along homebound, it is a fine sight to see these large clipper ships carrying double topsails, double topgallant sails, royals and skysails.

Sunday, November 7, 1880

This day came in with strong trade winds. As the wind freshened, we had to shorten sail. In the afternoon, I got John Brown and Frank and gave them their lessons in reading and writing. Afterwards Sam Hazzard and I got to talking about his folks and his girl back in Massachusetts and the letters he got from them. Then Sam asked about my folks. When I told him that I had seven brothers, one sister, and my father and mother, all living in Philadelphia, he was surprised and wondered why I got no mail from them. I said maybe I would get some letters in the next port. Sam felt kind of blue about it and asked if I did not have a girl. I told him that I had no time to think about a girl, for I had to work too long and too hard before I went to sea, and since then had been on shore but five days. Sam said that he had seen me talking to a fine-looking girl in New Bedford the day before we sailed. She was a Scotch girl I met at dinner at Squire Butts'. As I only got in to New Bedford that morning and had only seen the girl for about a half hour, we had little time to get acquainted. I told Sam that my first

TO TRISTAN DA CUNHA 41

voyage had been up in Greenland for twenty-two months, and could get no letters, and before the second voyage north I had been on shore only three days, and when I got back from that trip, I left the same day for a trip to Russia.

Saturday, November 13, 1880

Ever since last Sunday we have had fine weather, light trade winds and all sails set. We have been busy patching sails and rigging. We took down the main topgallant sail, patched it, sent it aloft again and set it. We bent the fore topgallant sail, lowered it on deck, resewed it and put on cross bands to strengthen it. New pennants were put on the mizzen topmast staysail, and new sheets and halyards. We got a strong topsail out of the fore hold sail pen, put strong bands on it, and after taking down the fore topsail put this stronger one in its place. A heavy storm sail was restitched, also the bolt rope around the clew, the sheet, the tack, and the head, and stowed it down where it will be handy. We set up the topmast rigging and put on a few ratlines and some chafing gear in wake of the yards.

Lately there have been a few tropic birds around. They are pretty little birds, all white, with two long thin feathers in their tails.

Sunday, November 14, 1880

We were going along at a good clip this morning when Mr. Young, at masthead, sung out with a ringing sound that brought all hands to their feet and up on deck, "There she blows, a school of sperm whales!" The captain shouted, "Haul aback the main yard. Hoist and swing and when all's clear, lower away all four boats." The cooper took charge of the deck and the captain went aloft on lookout. At the first rising, Mr. Young got fast to a whale. Shortly after that, all of the boats were fast to whales. Then the fun began. Flukes were flying. The mate's whale rolled over on its back and snapped its enormous jaw here and there, so they had to slack line and give it sea room. The whale came for one boat after another. When it came at our boat, Mr. Gifford hollered to give it a bomb. Mr. Reed shot a bomb down its throat when it was only fifteen feet away. In about three seconds, we heard that bomb burst. The whale shot forwards toward our boat and as I swung the boat over, it heeled over alongside. Mr. Young helped lance our whale. The fourth mate was having trouble about a quarter of a mile away and calling for help. His waif was flying. Mr. Gifford told Mr. Reed to go to Mr. King and see what his trouble was, while he would get our whales to the ship, which was only a couple of hundred yards off. When we got to Mr. King, his boat was stove. One of his men had put his shirt in the hole and the men had to bail fast to keep the boat from filling. Their whale was dead, so we towed the whale to the ship and sent the stove boat ahead. We were busy cutting in all night. Half the watch went below for rest, and the watch on deck cleaned up the falls and the cutting gear.

Thursday, November 18, 1880

We finished boiling oil yesterday, smoking like a steamer, with all hands dirty and greasy, but that is what we want and plenty of it. As the oil was too hot to stow below, the casks had to be lashed on deck.

Today we stowed the oil down in the after hold, seventy-four barrels. Everyone was tired, hungry, and dirty, but we had to clean up decks just the same. Mr. Young, Joe King, Sam Hazzard, and Manuel Gomez are all on the sick list.

Sunday, November 21, 1880

Saw two ships at sunrise this morning, one bound south around the Horn and the other close by, bound north. The latter looks as if she had had a hard time of it. Her fore and main topgallant masts were both carried away and the jib-boom was broken off short at the collar. She must have been caught quick and sudden when she was coming around the Horn. Her boats and bulwarks were smashed, just the bare stanchions standing. Her crew was busy fixing things, trying to clear the tangled rigging. The men on deck were getting spare spars ready to put up aloft. She was very deep in the water and must have strained her seams.

Thursday,. November 25, 1880

A good breeze from the north and all sails set. We sighted a whaling barque and ran down close to her. The captain lowered a boat and went on board of her for a gam. She was the *Pioneer* of New Bedford, a high and short barque with no sheer, a square bow, and square stern. She looks as if she were built by the mile and cut off any length, but they say she is a comfortable old tub. Today is Thanksgiving Day but we had no turkey. Instead we had one of the little pigs for dinner. The pig tasted good with roast sweet potatoes. All of our ducks and hens were gone long ago.

Sunday, November 28, 1880

We got two porpoise today and had porpoise balls fried with onions for supper. They are quite a treat for a change from salt horse and salt pig. This afternoon it got to breezing up strong and we let go sheets and tacks, clewed up the courses and furled them. We had to reef both topsails, haul down the jib, and brail in the spanker. The main and mizzen staysails were hauled down and made all snug. After dark there was a gale of wind and it got pretty rough. Heavy seas came over the rail, so we hoisted the boats on the upper cranes and put on extra gripes. We took the line tubs out of the boats. The main deck was full of water, swishing all about. Everything loose had to be lashed and all of the hatches battened down.

Tuesday, November 30, 1880

The gale is abating but it is still very rough. On the lookout up aloft, it would shake the insides out of you when the ship plunges into a head sea. There was plenty of dinner for the fish today. I saw nothing today but some Mother Carey's chickens following the ship. They are pretty little birds, black and white and awful quick and playful. They are often seen following a ship when it is stormy.

Wednesday, December 1, 1880

It was very wet this morning, but seeing a school of blackfish we hauled aback and lowered four boats. We got three and the boy hooked a fine albacore. The blackfish were hoisted on board one at a time with a large tackle. They were of good size, weighing from fifteen hundred to two thousand pounds. The blubber was skinned off and the carcasses thrown overboard.

Friday, December 10, 1880

With the exception of a couple of days, the weather has been stormy since Thanksgiving. Yesterday we gammed the whaling barque *Andrew Hicks* and today the barque *Petrel* of New Bedford. Manuel King sighted a right whale. Although the sea was rough, we lowered three boats and, together with the boats of the *Petrel*, soon killed it. The *Petrel* came alongside and started to cut in with some of our crew to help. We got half the oil and half of the bone, amounting to forty-two barrels of oil and four hundred bundles of bone.

Thursday, December 16, 1880

It is thick foggy weather. There are lots of "gonies" (albatross) around the ship today. They are large birds, about the size of a goose, and have webbed feet. They have large hooked bills. Their wing spread is enormous, at least six or eight feet. They are seldom seen north of the Equator, only under the Southern Cross and mostly between the two Capes. If they should be caught and left on deck, they could not fly away. I think the suction keeps them down. When the ship is lying idle, they will come right up to you and take morsels of food from your hand. These big birds do not seem to have any fear and look at you with their innocent eyes, not bold, but with a childlike trust. Sailors are very superstitious about these birds and would not hurt or kill one for anything.

FROM TRISTAN DA CUNHA TO ST. HELENA

Getting supplies from Tristan da Cunha—Christmas Day—Right whales—Scrimshaw work—Oar rammed down whale's throat—Sick man taken to St. Helena—Letters—Supply schooner—Transferring oil—The Seaforth Highlanders—Church services—Dinner with the Port Officer—Scotch lassies—Treated like a son—A sovereign earned—Reminiscences of Scotland—Jamestown barracks—Off for the Indian Ocean.

CHAPTER III

FROM TRISTAN DA CUNHA TO ST. HELENA

Saturday, December 18, 1880
Today we sighted Tristan da Cunha, a group of small islands about Lat. 37° 7′ S and Long. 12° 2′ W. These high and rocky islands are very dangerous for ships at night as there is no lighthouse on any of them. It is very hard to make a landing. The water is very deep and the rock almost perpendicular. There is a small vessel sent by the British government that comes here twice a year to bring supplies. The people raise cattle, sheep, goats, and plenty of potatoes, onions, and other vegetables, but no fruit except apples. The houses are built of stone with the roofs thatched with straw. Because of the terrible storms that they have here, the houses are built very strong. It never gets very cold. The people make their clothing from the wool of the sheep. There are about 250 people, men, women, and children, living on the largest island in a large village. They are a rugged, strong, and sturdy lot and not bad looking. They are very healthy and have little illness. Their furniture is all homemade but comfortable. There is a school that does for a meetinghouse. The schoolmaster is minister, doctor, judge, and head man of all affairs. They have several big strong boats for fishing and for use between the islands. They catch lots of fine large cod and other fish. Cattle are shipped to the island of St. Helena. Their winter, although there is no snow, comes in the months of June and July. Sometimes it is dangerous to leave the house, the wind is that strong, coming in sudden gusts from the ice fields lying to the south. The people are mostly of Scotch descent but the Scotch language has died out; only once in a while they talk broad. They live to a good old age, some reaching one hundred and ten. They have no police and no jail and do not need any as they are good people.

    I went on shore with the captain. We could not land the boat. There was no beach and the men had to keep it back from the rocks. The people greeted us kindly. The captain bought five bags of potatoes and three bags of onions. He bought a hind quarter of beef and they dressed two sheep for us. The boat had to make three trips to the ship but it was only a short distance. Each time I had to wade out to my armpits with the bags and mutton on my back. The hind quarter of beef was heavy and the rocks very slippery. The captain paid for the supplies with calico, sailcloth, whale rope, and tobacco. They were very pleased to get these things, especially the whale rope. When they bade us good-bye, they gave us a basket full of good fresh eggs, about twelve dozen.

Saturday, December 25, 1880

Christmas. Had a leg of roast mutton, baked potatoes, both sweet and white, onions, baked beans, and pie for dinner. This was a dull and dreary day, rather rough weather. Everybody paced the deck and talked of home and the good things of a New England Christmas dinner. Mr. Young was wishing he was back home in the Sandwich Islands (Hawaii). Mr. Reed was wishing he was in little old St. Helena. Then Mr. Reed asked me where I would like to be. Before I could answer, Sam Hazzard spoke up and said I would like to be back on the island of Flores with the pretty barefooted Portuguese girl who kissed me good-bye in front of the priest. Everybody laughed. I said I was well satisfied with my lot if we could only see some whales. Mr. Reed said, "Wait till I get you in St. Helena and I'll let you see some fine girls there." I said nothing but thought to myself of places where I had been, like the West Indies, and of some of the fine girls I had met there.

Friday, December 31, 1880

Yesterday we saw some humpback whales, lowered the boats and gave chase but got nothing. We gammed the whaling barque *Hercules* in the afternoon. This morning about eleven o'clock, Mr. Young sung out good and loud, "There she blows and there she blows!" and aroused all hands in a jiffy. Mr. Gifford cried, "Where away?" Mr. Young answered, "Down on the lee beam about a mile off." There were two whales that seemed to be feeding. There went their flukes. They were right whales. Mr.

Gifford told Mr. Young to come down from the masthead and get the boats ready. The captain went aloft. After shoving clear of the ship, we set masts and sails and ran down to the two whales that were quietly lying there like logs. Before we got near, they turned up flukes. All three boats lay around about an eighth of a mile apart waiting for them to come up. Finally they both broke water ahead of us and near the mate's boat. Mr. Reed hauled aft his sheet and headed straight for the whales with the mate coming up behind. We got to the whales about the same time. Mr. Gifford kept to the leeward and we had to luff up and go to windward. That gave me a left-handed dart but Mr. Reed put me on top of the whale so I got both of my irons into it. It was all we could do to keep clear of the line, it ran out so fast. Then the whale slacked line and when it came close, Mr. Young lanced it. The mate killed his whale in the meantime. We hauled line up to our whale and tried to lance it. Mr. Young's boat got stove and had to go back to the ship, the men using their coats to keep the water from pouring in.

I put Mr. Reed up close to ours and he kept jabbing a lance into it until the boys commenced to holler that it was spouting blood. "Stern all, stern all!" Mr. Reed cried. None too soon as the whale went into a desperate flurry, thrashing the water with its tail, diving and breaching until it finally rolled over on its side, dead. These whales gave us 140 barrels of oil and a lot of bone.

Saturday, January 22, 1881

Things have been pretty dull for nearly three weeks. Got one blackfish on the eleventh, one on the fifteenth, another on the sixteenth, and four on the nineteenth, but no whales. The officers and harpooners are all starting to scrimshaw, that is, to make little things out of ivory, rosewood, and ebony. There are lots of things of this sort hanging around the vise bench where all the men gather to work. We caught some fine albacore today. Fresh fish are welcome to our daily fare. The albacore is a fine fish with only a backbone, and the meat is like a Spanish mackerel.

We killed a big shark and hoisted it aboard to get enough pieces of skin to use for sandpaper. It is fine for smoothing down ivory and cuts just like emery paper, but it works best when wet.

Wednesday, February 2, 1881

Set sail at daylight with the weather warm and the trade winds steady. Just at breakfast the welcome cry rang out from masthead, "There she breaches, again she breaches. There she whitewaters and there she blows and there she blows!"

"Where away?" the captain cried.

"Five miles astern," was the reply.

"Up your wheel, brace around the yards, finish your breakfasts. Sing out when we head for them."

When masthead sung out again, Mr. Reed ordered, "Steady right ahead, let her luff two points. Come down and get the boats ready. Cooper, you go aloft on the lookout."

All hands were anxious and ready to jump into the boats, and the ship keepers were at their stations ready to man the braces. When about three-quarters of a mile from the whales, we made for the school.

The mate had a big new lugsail, the second mate an old gaff sail, the fourth mate an old spritsail, and we in Mr. Reed's boat had our new topsail set. The wind being a little strong, we had full and plenty with the four Portuguese sitting on the weather gunwale, their paddles in their hands ready and eager for the word. I was standing up in the bow looking all around, when one of the Portuguese said, "The whale is up." He got the signal from the ship which was watching them from the distance. The whales went down again, so we all slacked sheet and waited for them to come up. They did not stop down long. All four boats rushed to the middle of the school and got fast to the whales about the same time. I changed ends to steer and Mr. Reed stood ready with his lance, watching for a chance. The whales were running here and there, heading for a boat one minute and letting their flukes fly. The next minute they would be sounding, and the next breaching half out of water. The fourth mate had two oars broken and his boat stove. We had a narrow escape

TO ST. HELENA 51

when one fellow came at us with its jaw extended up in the air and aimed straight for the middle of the boat. Frank Gomez rammed his oar down the whale's throat and pushed hard and choked it. It chewed the oar but left the blade sticking in its throat and then went into a flurry and drowned itself. Mr. Gifford's whale ran to the outside of the school and he killed it easily. Mr. Young had some fun with his whale, which kept its tail going like a windmill, and he had to wait until it was tired out. We finally got all four whales to the ship about three o'clock. It took until midnight to get them all cut in. One hundred and forty-five barrels of oil were stowed away from this catch.

ST. HELENA

Thursday, February 10, 1881

The past few days we have been going over our sails, patching and strengthening them and generally looking over the rigging to see if it is in safe condition. Today we sighted the island of St. Helena ahead and reached the harbor of Jamestown at three o'clock in the afternoon. The captain went on shore and sent the boat back to the ship with the mail. I received seven letters, the first since leaving home nearly a year ago. One of our men, the cooper's brother, who has been sick for such a long time, was sent on shore. He is very bad and looks as if it was all done with him. He has the consumption.

We arrived here sooner than expected, so the men were disappointed because they did not get many letters as most all of the mail is coming on the supply schooner now on her way from New Bedford. She may not get here for a month yet. The captain will stay on shore until tomorrow while the ship is laying off and on across the harbor under short sail. The main reason we stopped here is to put the sick man ashore.

Tuesday, February 15, 1881

We never got to go ashore at St. Helena. The men gazed with wistful eyes at the island. Some of the officers were just as anxious to go on shore, especially Mr. Reed, who has a girl there. The island looked so grand and inviting after the storms around the Cape and the Tristans, but as it will be a month before the supply ship will come, we did not anchor and went on our way. Tonight was splendid moonlight. The glimmer of the moon on the water and Tommy Wilson's singing "Rolling Home to Dear Old England" made me feel kind of funny. This little Englishman has a rich baritone voice. He is a man about thirty years old, not stout, rather thin, and a good shipmate and true.

Sunday, February 27, 1881

There has not been much to do lately. Some of the men have been doing scrimshaw work and making canes. Others have been making gimcracks of ivory, paper knives, cribbage boards, ditty boxes to hold buttons, needles, thimbles, scissors, spools of thread and things for sewing. Some have been patching their clothes. I am making a cap with a good-sized peak or sunshade for the eyes. It seems strange to see the men making clothes, light duck breeches, jumpers and even canvas shoes, and making good jobs of them.

A few days ago we gammed the full-rigged ship *Eliza Adams* of New Bedford, a fine-looking four-hundred-ton whaler, and always considered lucky. A day or so later we gammed the whaleship *Jerry Perry*, a regular clipper, fine lines and low in the water, with her prow and spars all painted black.

Tuesday, March 1, 1881

Whenever I could find the time, I have been giving Frank Gomez and John Brown their lessons. Frank pulls an oar for me and is a very good fellow in a boat. He is very anxious to learn to read and write. To offset the lessons, he does little odd jobs for me, like washing my clothes and helping me out at any time. Down in the boat when pulling to windward for five or six miles, if I ask, "Tired, Frank?" he will reply, "Not much," and laugh. John Brown is a negro and very black, but he is very much of a man and quick to learn. He pulls the midship oar in Mr. Young's boat.

TO ST. HELENA

Saturday, March 5, 1881

This forenoon we sighted a large mail boat and passenger steamer, one of the Donald Currie line, bound south to the Cape of Good Hope. Her deck was swarming with passengers watching us. There were soldiers on board and the band was playing, "The Bonnie, Bonnie Banks of Loch Lomond." We dipped the Stars and Stripes. Up went a ball to their main truck and out burst the British flag. On a staff at her stern was the royal banner, denoting a naval officer in charge. Then the band played "Dixie."

Sunday, March 6, 1881

Everybody is writing letters, answering those received at St. Helena. We are all expecting to be back there before long, when the supply boat comes. I wrote to one of my brothers who is on the *S. S. Pennsylvania* of the American Line. Sam Hazzard has written five or six letters and some of them bulky. I don't know what he finds to say or is it half nonsense he has been writing all week whenever he had time. Frank Jose asked me to write a letter for him to his girl in New Bedford and some of the things he wanted in it were laughable. Then Frank Gomez wanted me to write to his girl in Flores, and as it was a pretty decent letter Frank said, "It is very much good." I asked him why he did not write to his mother and he said, "All same. My girl will tell my peoples all about me."

Monday, March 14, 1881

Not a spout for a week. Mr. Reed got a nice board and asked if I would paint a name on it for our boat. I cut the wood into a neat shape, smoothed it off thin, marked the name *Mary Joshway* on it and placed a star at each end, cutting out the stars with a chisel. Mr. Reed wanted the name cut out too and I did that with my knife. I painted the board white and the letters dark green. As I held the board in my hand, the captain came along and asked if that was the name of my girl. I said that was the name of Mr. Reed's girl in St. Helena. The officers laughed and Mr. Reed turned scarlet. Some of the boys who had been in St. Helena knew this girl and told me that she was a half-caste and not at all good looking. Mr. Young said, "Bob, you'd better ask Sam Hazzard if he does not want the name of his girl on our boat." Sam got a fine piece of rosewood and I cut out the letters with a good sharp chisel, sandpapered it with shark skin and rubbed it with beeswax until it got hot and shone like glass. The letters that were cut out I painted white, and all were pleased with the job. The captain called me aft and asked me if I would paint his cabin the next time he went on shore. I told him that I would do the best I could.

ST. HELENA

Wednesday, March 16, 1881

At daylight we sighted the barque *Morning Star* and the island of St. Helena close by, but we kept off and around the eastern end of the island until about one o'clock when we were off Jamestown. The captain lowered the starboard boat, went on shore and left word to lay off until morning. We got the cable, shackled it to the port anchor, overhauled a range of chain and got the anchor all ready to let go. Everybody was talking tonight about St. Helena and anxious to get on shore. Tommy Wilson did not feel like sleeping and sang, "Beautiful Isle of The Sea" for us.

Thursday, March 17, 1881

After breakfast we lowered the starboard boat and brought the captain on board. We took the ship in the anchorage at eleven o'clock, took in all light sails and furled them, hauled aback and let go anchor in thirty fathoms of water, slacking out about seventy fathoms of chain. The running rigging on the tops and sheer poles was coiled up out of the way and off the decks. We got the fish tackles over the main and after hatches, to break out the oil to ship home. Next we got out slings, cant hooks, and small handy tackles.

Nearby was a small whaling barque *Clarissa*; a little further the *Charles W. Morgan* and nearby the schooner *Franklin*.

TO ST. HELENA

Friday, March 18, 1881

There were lots of fish, large and small, and plenty of Spanish mackerel all around the ship. After breakfast came the call to break out oil from the main and after hatches. Some was sent aboard the *Charles W. Morgan*, which belongs to the same owners as the *Kathleen*. Our supply ship, the three-masted schooner *Lottie Beard* of New Bedford, came in and anchored. Several large British ships are here for orders and water, the latter being brought out in large boats built for the purpose. At five o'clock the starboard watch, including Sam Hazzard, Mr. King, Mr. Reed, and Mr. Young, went on shore. Some of the rest of the men went gamming on the *Charles W. Morgan*, to get news of old friends in New Bedford.

Saturday, March 19, 1881.

Busy all day until four breaking out oil. At five o'clock the port watch went on shore, passing the whaling barque *Wanderer* coming in.

This is a busy little place on account of so many whaling ships coming in to recruit. There is one wide street running up through the center of town with two or three cross streets and two more streets at the foot of the rocks which rise almost perpendicularly in some places. Most of the houses are small, built of stone and very strong. Five or six families cook in one kitchen, built at the back of the houses. The water is very good, nice and cool, and comes down in the main street from away up on top of the rocks.

There are two churches, one Episcopal at the foot of the street, and up at the head of the street is a one-story Catholic chapel built and supported by the government for the benefit of the soldiers. On the side streets are several fine houses. For a small place like this, there are plenty of grog shops, few stores, and a bakery and coffee shop combined. I had to hurry to get on board when I heard the warning gun boom, a warning to get through the gates, which close at ten o'clock. A sentry is stationed at the gates night and day.

Sunday, March 20, 1881

All the officers except the chief mate, Mr. Gifford, and one half of the men went on shore. I did not get to go and stayed on board listening to the church bells ringing and the bagpipes screeching up at the Fôrt. I could see the Highlanders marching down the hill to the church as I viewed them from masthead with the glasses. Later, I watched the people walking on the promenade. After dusk somebody on one of the whalers was playing a violin while others were singing.

Monday, March 21, 1881

We are still breaking out oil, and the cooper is kept on the jump tightening casks to ship home and setting up casks to stow down below. We shipped all of our oil home on the *Charles W. Morgan*, 460 barrels, and eighteen hundred bundles of whalebone.

The barques *Hercules* and *Stafford* came in and anchored this afternoon. At four o'clock we knocked off work and I went on shore with the port watch. I looked around for a place of amusement or a dance hall but found none, so I walked around by myself taking observations. I strolled down to the garden and up on the promenade. I could see the ships lying at anchor with their riding lights hung in the fore rigging, and could hear a Portuguese guitar or accordion squeaking across the water. Boom went the gun and as I was going through the gates I heard one of the guards address another in Gaelic. I turned quickly and answered him for I thought he was speaking to me. He held out his hand. I grasped it. I had to sit down in the guardhouse and talk with several Highlanders. They belong to the Seaforth Highlanders and asked me to come up to the barracks. It was ten-thirty before I got to the gates but the sentry let me through to go on board.

Tuesday, March 22, 1881

The cooper and two helpers were still quite busy driving hoops and setting up casks, tightening them and restowing the fore hold. Part of the crew under the charge of Mr. King were busy getting firewood from the condemned whaler *Draco*. Today the whaling barques *Falcon* and *Morning Star* came in to anchor. The mail steamer *Balmoral Castle* arrived also. She plies between Liverpool and South Africa. All of the men are happy and having a good time here. The starboard watch is on shore tonight. Tommy Wilson stayed on board. I told him about the Highlanders I met at the gate and he told me to keep in with those fellows as they often have a dance up at the barracks on Ladder Hill, and that I might have a chance to go although they don't often ask civilians.

Saturday, March 26, 1881

I have had no chance to go on shore since Monday. Since then we have stowed down two hundred barrels of fresh water and restowed the holds. Received provisions sent to us by the supply schooner *Lottie Beard*; coffee, tea, sugar, rice, barley, pickles, raisins, and molasses. Some of the officers got letters from home. Yesterday morning the sick man we sent ashore a month ago died.

The men have been catching a lot of small fish and Spanish mackerel alongside. The fresh fish taste good.

Sunday, March 27, 1881

One half of the crew go on shore today to stay until Monday morning if they want to. I shaved, dressed up, and went on shore just as the church bells were ringing. The door of the church was wide open so I walked in and just got seated when the Highlanders came marching up the aisle, playing the pipes. This was the Episcopal church and there was a good sermon by the young assistant to the minister, who was away.

When I came out and was standing watching the soldiers forming into line a man came up and spoke in a friendly way. He was Mr. Jamieson, one of the church officials and also Port Officer. He had noticed that I was a stranger. When he found out that I was Scotch, he asked me to go up to his house and have a bite of dinner. He would not take no for an answer and locked his arm in mine. I told him that I was on board one of the American whalers and that my people were Scotch. He said that he was too and moreover talked rather broad. He had a large house and a nice garden full of roses and beautiful flowers. He introduced his wife and six daughters. We had a very fine dinner. His wife spoke like my mother, very broad. She said she came from Paisley and when she learned that my folks belonged there and that my grandfather lived on Glen Street and my uncle on Caledonia Street, she would not let me leave the house but kept me sitting there talking. She told one of the girls to bring Lady Ross. The oldest daughter, a nice girl of about twenty, soon brought back Lady Ross, a fine old lady about sixty years of age. When Mr. Jamieson introduced me to her, her first words were in Gaelic. It made me feel good to hear the tongue of the Western Highlands again. Lady Ross said that I would have to go back to her house for tea and talk about the Highlands. From the way that I spoke Gaelic, she said that I must come from some place in Rosshire. I told her that I came from Ullapool in Loch Broom. She came from Greenyard. I mentioned Major Duff to her and she said that he was her neighbor.

At five o'clock we walked up to the head of the street, and one of the Jamieson girls walked with us. Lady Ross has a fine house here and keeps two servants although she lives alone. As we were having tea, she asked how long I was going to stop on shore. I told her I would have to go on board tonight. She said, "Why not stop until morning?" and after talking, encouraged me to stop all night. At ten o'clock she told me to see the young lady home and be sure to come back. I started down the street with Miss Jamieson. She was very pleasant. She talked very broad, just like her mother, and said, "Lady Ross has taken a great liking to you, Mr. Ferguson. She is a fine old lady." When I got to her home she asked when I would be on shore again. I was not sure but told her most likely on Tuesday night. After bidding her good night, I went back to Lady Ross. We sat and talked about the Highlands and places that we both knew, talking mostly in Gaelic. We had a pleasant evening and she said, "I am going to call you Robert." At eleven o'clock she opened the large Bible and asked me to read a chapter. This Bible was just like the one my mother brought from Scotland. Then she said a Gaelic prayer, lit the candle and called one of the servants to show me to a room and bade me good night. In this fine clean soft bed I did not fall asleep for some time, but lay thinking of the kind way Lady Ross received me and how two days ago I knew no one on the island.

Monday, March 28, 1881

This morning I awoke, washed, and came downstairs. Lady Ross was waiting for me. For breakfast I had a plate of porridge with goat's milk, bacon and eggs and fruit. She said a real old-time Gaelic grace that did me good to hear. I did not hesitate to tell her so. Then she said to me, "While your ship is waiting here, Robert, I want you to consider this your home. It gives me great pleasure to have you here, to talk the dear old language and to think that we come from or near the same old Highland home. Don't forget to come here when on shore again as I'll be expecting you."

I hurried down the length of the town and out on the jetty I found Mr. Gifford standing there waiting for our boat. He told me to take a boat and go over to the packet ship *Milton* to see the mate, Mr. Conley, who wanted the name painted on the stern.

Tuesday, March 29, 1881

I got up at five o'clock and went over to the *Milton*. She is an old packet ship of over four hundred tons. Her windlass is abaft the foremast, the only one I ever saw there in all my travels. Mr. Conley had me down for breakfast with him in the cabin. I finished painting at noon and went back to the *Kathleen*, shaved, dressed, and cleaned up. The mate called a boat's crew and told me to jump in as he was going on shore. I took a walk around town and had a look at it in daylight. As I was going down toward the coffee house, one of the boat's crew told me that the captain of the *Milton* wanted to see me. I met him and his wife. He thanked me for painting the name on the stern and gave me a sovereign.

While talking to Captain Potter, one of the Jamieson girls came along. I spoke to her and went down to her house. They were all glad to see me. They helped me to a good dinner at two o'clock. I was hungry. Afterwards I had a talk with Mrs. Jamieson and her daughters, of places around and about Paisley in Scotland, such as the Red Sniddy, Jail Oppen, River Cart, Coates Mill, High Street, Moss Street, Snedden, Brig, Canal, Hope Temple Garden, Johnston, Renfrew Road, the race courses, the Infirmary, the Greenock Road, the large elm trees and the starch works in Caledonia Street. When the dinner dishes were cleared away, the girls sang and played some of the old Scotch songs on the piano and the violin. Two of the girls sang "Loch Lomond" and their mother sang "Mary of Argyle." Mr. Jamieson came in and sang "Annie Laurie." After the music I said that I would have to go and call on Lady Ross but the girls wanted me to tell them all about whaling so it was after supper before I got started. Two of the girls and I took a walk in the Governor's garden and then walked up the street to Lady Ross' house where I was going to stay all night.

The old lady was glad to see us. The girls came in and we had a fine time. I certainly had a pleasant day and feel thankful to Lady Ross. No mother could have been more considerate. With her quiet Highland ways, she made me feel quite at home.

Wednesday, March 30, 1881

After a good breakfast Lady Ross said, "Be sure and come up whenever you can." I got on board at nine o'clock and changed clothes for my dungarees. Got stores out of the after hold and cleaned ship. This is our last night at anchor for a while. All hands like this place. As for me, I think of the good times I have had, yet knew no one when we dropped anchor. I took another look at this lovely clean place, sitting in a valley between big hills. The rocks rise perpendicularly from the water's edge on each side of the valley. Between these, a wall at least forty feet high has been built from side to side, forming a promenade about seventy-five feet wide at the top. There is a large oaken and iron gate opening to the jetty or landing which runs obliquely along the foot of the rocks. On the western side, the rocks run up for a thousand feet to a large rocky platform, eight hundred feet long and six hundred feet wide. The barracks are located up there, making very pleasant quarters. There are several large and heavy guns up on this plateau. From the foot all the way to the top are steps cut in the solid rock, some seven hundred and fifty. There is another road that runs zigzag for the soldiers to march up and down. On the other side is a masked battery cut deep in the rocks halfway up the breast of the rock. At the foot of the promenade is the Governor's garden, containing some large trees with benches under them, fine shrubbery and flowers. Away up at the top of the street is the rough shady road leading up to Napoleon's tomb and Longwood.

One side of the valley is all green, covered with cactus or prickly pear, and the other side is one mass of red and pink geraniums all in bloom.

Thursday, March 31, 1881

After breakfast we loosened the gaskets of the topsails, got the rigging down on deck, loosed all sails ready to hoist, hove short on the anchor, and hoisted fore and main topsails with the chantey, "Good-bye, fare you well."

We set the fore staysail, hoisted the jib and hove up the anchor, keeping off before the wind. After we got the anchor up to the cathead, we set fore and main topgallant sails, hauled out the spanker, got the anchor inboard and lashed, unshackled the chain and stowed it down below in the chain locker. The captain did not come aboard so we lay off and on. The boys feel sorry at leaving this island for they all have had a good time at small cost. Mr. Gifford gammed with the whaling barque *Bertha*. Tommy Wilson, Sam Hazzard, and I were talking. Sam asked how much money I had spent. I said about one shilling and that was for the sermon I listened to. Tommy spent one pound and Sam, three. They wanted to know if I spent the night at the barracks. I said, "No," and then Sam said, "Leave it to Bob. I'll bet he had a skirt on the string." Well, what they don't know won't hurt them.

FROM ST. HELENA TO COMORO, SEYCHELLES, JAVA, CELEBES, AND BACK TO COMORO

Off the Cape of Good Hope—Bad weather—The Flying Dutchman—A monster sperm whale—Believed on fire—Monsoons—On shore at Europa Rocks—Turtles, birds, eggs, fruit—Making trinkets—Sponge cake—Zambesi River—Comoro Islands—Sour-dough bread—Bird Island—Eggs and fish—Seychelles Islands—In company with the Osceola—Two-legged pigs—Christmas Island—Battle with Malays at Lombok Straits—Squalls—Hot weather—Osceola loses her mizzen mast—Macassar, Celebes—Sunda Straits—Trading at Aldabra Islands.

CHAPTER IV

FROM ST. HELENA TO COMORO, SEYCHELLES, JAVA, CELEBES, AND BACK TO COMORO

Friday, April 1, 1881

After the captain came on board this afternoon, we were off on the port tack with every rag on her, setting a course to the southeast. Some of the men are superstitious about sailing on a Friday and April Fool's Day too. All the whalers left the harbor today and will scatter over the South Atlantic from the Rio Plata to the Congo and from the Tristans to the equator, searching for whales.

Saturday, April 9, 1881

We have been having fine weather ever since leaving St. Helena, but have had little to do except go over the sails and the rigging, patching and strengthening them. Our course is still southeasterly and we find the weather getting much cooler the further south we go. The men going aloft wear their coats to keep warm and on deck they don't sit around any more. Some of them are wearing mittens and sea boots in which they can hardly walk when there is a slant to the deck.

Sunday, April 10, 1881

The boys have been talking today about the good times they had at St. Helena. They have been rather curious about what I did. Mr. Young said, "From what I hear, and what I saw, Bob had as good a time as any of you, although a stranger on the island. I saw him one night with two fine-looking dames in the Governor's garden, but he did not see me for he was too attentive to the light-haired one, and believe me, they were the pick of the town."

Friday, April 15, 1881

The weather is very rough and wet. There's a big swell on, with water sloshing across the deck, slopping water over the lee rail all the time and the ship rocking like a cradle. Yesterday we took in some sail but today had to shorten more and finally hove to with a heavy sea running. As we were shipping tons of water, we hoisted the boats on the upper cranes and took the line tubs out. We stood in water on the main deck up to our thighs at times. All of the hatches were battened down. Life lines were strung along the deck from the windlass to the break of the poop and aft to the taffrail. Everybody was soaking wet. Oilskins were no good. Everything was double lashed and even the topsails were lashed to the yards. Watch tackles were left here and there, handy in case of need. The wheel was lashed and all ready to be cast off in a moment. Only the officers and harpooners are on deck tonight, keeping a wide-awake lookout. It is a dark and stormy night.

Sunday, April 17, 1881

The gale has abated some but the seas are still high. We set some light sails to steady the ship. About three o'clock John spied a school of sperm whales. We lowered four boats. Mr. Gifford and the second mate each got one. By four, we had them all alongside and the fluke chains on. The water was very rough. We cut in all night for fear of a worse storm.

Thursday, April 21, 1881

On account of the high seas running, it has been slow boiling. All blubber had to be taken off deck and stowed down the main hatch in the blubber room. The hot oil has to be carefully watched so that it does not slop over. All casks on deck filled from the cooler had to be lashed. However, today we finished and stowed 120 barrels of oil.

Sunday, April 24, 1881

We're having disagreeable weather with rain, squalls, and fog banks drifting up from the south. The boys amidships are talking about the *Flying Dutchman*, each giving his version of this mysterious ghost ship which is supposed to be seen down here off the stormy Cape of Good Hope and stated to have been seen and sworn to by many sailors. While enveloped in a thick

JAVA, CELEBES, AND BACK TO COMORO

fog bank this afternoon, four men and the third mate sang out, "Sail ahoy, right ahead and coming right for us!" She had a tremendous spread of canvas. The mate standing aft next to the captain sang out, "Up with the wheel, lively," and he sprang to help the helmsman. A large full-rigged ship passed not a hundred feet from us. She was painted all white with a thin black stripe and looked like a ghost in the fog. It was the famous Yankee clipper ship, *White Cloud,* a noted fast sailer. If she had struck us, we would have crumpled all up. After this happened, the men all told yarns about other narrow escapes they had. It made us all more careful when on watch in foggy weather.

Sunday, May 1, 1881

Except for finbacks and a few porpoise, we have seen nothing for a week. Today my thoughts wandered back home to my mother and her many kind acts toward me. This is unusual for I am never homesick. It is calm today, the sails hanging around the masts and the yards, slatting as the ship rolls in the swell. The men were reading books and papers, cutting one another's hair and shaving. Some were talking of their folks or their girls or telling of their experiences.

Friday, May 13, 1881

The weather has been bad, rough, rainy, and squally for several days. The seas have been so high that nobody has been at masthead for it would shake the insides out of you up there. We are past the Cape and heading in a more northerly direction. Today two of the men down below forward had a battle royal. The mate fetched them on deck, lashed both of their left hands together and gave each a piece of rope about three feet long and told them to lay on to one another. They did lay on for a while but soon got all the fight that they wanted.

Monday, May 16, 1881

I was up at masthead at daylight and at seven o'clock sang out, "There she blows. Sperm whales on the port bow!" At the second rising, Mr. Gifford got fast, but the whale immediately rolled his boat over, upside down. We pulled up to the mate's boat and set it right again but Mr. Gifford climbed into our boat. During a battle of flying flukes and snapping jaws, we lanced the whale and finally killed it. This was a good whale, giving us ninety barrels of oil.

Sunday, May 22, 1881

I was at masthead until breakfast time. Just as the sun came up large and round, I saw a whale about four miles off and sang out with a thrill, "There she blows, a lone whale right in the sun glade about four miles off!" Three boats were lowered and run to the eastward where they lay and waited for the whale to come up. Mr. Reed put me on to it and I got both irons into it good and solid. When I struck, the whale sounded and took nearly all the line out of the two tubs, at least eight hundred yards, and then came up with a breach clean out of water and a snort you could hear a mile away. It flopped down on the surface with a noise like thunder and splashed everybody in the boat. It put up an awful fight with its large jaw gaping and snapping at us and fanning its tail trying to kill us, but the boat's crew was too well schooled and knew when to back water. They were on the alert for my words, "Stern all." We kept out of the way of the lashing flukes and gnashing jaws. All the time Mr. Reed was waiting for a chance to lance it. After a long churning with the lance, the whale finally turned over and the three boats towed it to the ship only a short distance away. The head was too large and heavy to hoist in so we had to separate the junk from the case overboard and bail the case at the gangway.

Tuesday, May 24, 1881

Last night was very dark and a passing ship came alongside, lowered a boat and some of their men came aboard wanting to know what was the matter. They thought we were on fire. We were certainly firing up and there was a blaze going up from both stacks. One man in the boat was a bluff Englishman

JAVA, CELEBES, AND BACK TO COMORO 69

who said, "I thought that your ship was on fire and that there was the devil to pay." He bawled out, "Greasy luck to you," and laughed long and hearty. Mr. Gifford thanked him for his trouble and said it was very good of him to stop.

Thursday, May 26, 1881

We finished stowing down today, 120 barrels from this lone whale, the largest sperm whale caught in many years, so Mr. Gifford said. The teeth measured seven inches high, five inches broad, and three inches thick. The men all feel tired, dirty, and hungry. After supper we shortened sail, hauled aback and lay on the port tack until morning.

Monday, June 6, 1881

The weather is fine with good steady southeast winds. We are getting into the monsoons. These winds are similar to the trade winds in the Atlantic Ocean, except they change every six months and blow in the opposite direction while the trade winds always blow the one way. The monsoons blow six months from the southeast and six months from the northwest, and they are never too strong except at the change. The men are all talking about the coconuts at Madagascar, and hope that they will get on shore. They heard the captain talking about Europa Rocks and expect to see them about sunset.

Wednesday, June 8, 1881

The captain and mate are somewhat puzzled about some small group of islands that are not in sight yet, and think the chronometer is wrong. These islands are in Lat. 31° 21′ S. Long. 39° 36′ E. in the Mozambique Channel between Madagascar and the coast of Africa. At noon we saw low-lying land about ten miles on the weather beam but did not shorten sail.

EUROPA ROCKS

Friday, June 10, 1881

After breakfast we hauled aback the main yard and lowered the boats. We are in close to Europa Rocks. The crew went on shore to get some turtles and brought three back to the ship. They were so big that it took three strong men to turn them on their backs. We pulled the head of one out of the shell and chopped it off with an ax. From this one we got a bucket of eggs. The cook stewed some of the meat for dinner and it certainly was fine. I never ate anything like it. The head of the turtle was as big as a coconut. When these turtles snap at a piece of cordwood they almost bite it in two.

These islands are low and sandy with some coconut trees, bushes, and mummy apples (papayas). No matter how small an island is, there are always two or three coconut trees growing on it. Turtles come here by the hundreds, and birds are so plentiful they are like flies around a sugar hogshead. You can hardly put your foot down without tramping on them. Some of the boys got some fine shells, mostly cowries. After supper two boats went on shore. The ship is lying only about five hundred yards away. The water is covered with birds, divers, boobies, gulls, and frigate or man-of-war birds. The frigate bird is a beautiful black-and-white bird with a strong spread of wings and is as bold as an eagle. It will fight for anything in the air or water and is swift and quick as a flash on the wing. It lays its eggs on the sand and the sun hatches them out. There is no nest. The ground seems covered with eggs, so close are they laid together. The air is full of birds, squawking continuously, and diving for fish, screeching or whistling.

The crew brought back some mummy apples. They eat and look like cantaloupes and are about the same size. They grow on a small tree. We had plenty of turtle eggs for supper and they eat all right fried. The water around these islands is teeming with fish.

Saturday, June 11, 1881

It's a fine day, nice and warm. It feels good after the cold, stormy weather around the Cape. Mr. Reed and I went on shore to stretch our legs and took dinner with us. We got two washtubs full of fresh birds' eggs. Little turtles by the thousands were seen scampering down to the water as fast as they could go. Turtles crawl up on the beach at night and lay about a bucketful of eggs in the hot sand and cover them up. In the morning they leave again for the water. Turtle eggs are a little larger than a hen's egg but round and with a soft tough shell which won't crack or break, and they are very good eating.

We are living high these days and letting salt horse rest, with birds' eggs fried or boiled for breakfast, turtle stew for dinner or supper, sometimes flapjacks or cake made with birds' eggs or a rice pudding with plenty of eggs. Again we have fried fish or baked fish. In places like this, the old greasy whaler has the advantage over the merchant ships and the foremast hands live much better than those in the cabin of any big ship or tramp steamer. I have been thinking how comfortably a person could live on these islands.

Sunday, June 12, 1881

We killed another turtle today and cut it up for dinner. Fore and aft, these turtles measure four feet five inches by three feet seven inches, and one big one measured five feet three inches by four feet eight inches. These measurements are for the shell only. One of the turtles laid a bucketful of eggs on deck.

I held school for an hour today with Frank Gomez and John Brown. Frank is getting along fine. John can read some small words but the larger ones puzzle him, so he comes to me at dog watch and asks the meaning of them. I have to explain, but he has a good memory.

Wednesday, June 15, 1881

We are sailing along under easy sail, working north and nearer to the African coast. I am starting to make a dipper out of a green coconut. The shell cuts easily and will get hard when dry. I am cutting a wreath of thistles around it, making the shell smooth. A piece of ebony tipped with ivory will be used for a handle. I took a shilling and cut it up for silver rivets.

Thursday, June 16, 1881

The steward made some sponge cake today, using turtle eggs. It looked good and tasted good but was just like India rubber. The officers made fun of the cake. I had a talk with the steward and told him what was wrong. He should have used the yolks only, not the whites, because the latter is what forms the shells and the flippers and becomes tough in cooking. He asked me to show him. I mixed a batch and baked it. When he put it on the table, the mate said it certainly was a good cake. The steward said, "You can thank Bob for that."

Friday, June 17, 1881

We are off the island of Juan de Nova or St. Christopher in Lat. 17° 3′ S., Long. 42° 47′ E. Lots of the men are making small ditty boxes and other things of ivory, rosewood, or ebony. I am still working on my coconut dipper and have the handle riveted on with silver rivets. Now I have to wait until the shell gets hard so that it will take a good polish.

Saturday, June 18, 1881

Today we are in close to land, the coast of Africa, a rough-looking land. I went to masthead at ten this morning and with powerful glasses I could see a large river called the Zambesi, with vapor rising like smoke from the falls. Opposite Quellimane I could see the wild beasts from up aloft, rhinoceros and other wild animals back on the hills. On the lowlands I could see herds of cattle. Every now and then antelope could be seen flitting along the high ground, their heads erect as if on the alert for the foe. Some animals were too far away for me to distinguish what they were, even with powerful glasses. We threw our turtle overboard today, thinking it was dead, but how it did scud when it struck the water. The weather is getting warmer and warmer as we go north.

Wednesday, June 22, 1881

We sighted several islands today, Mayotta, Mohilla, Johanna, and Big Comoro. These islands are known as the Comoro group

JAVA, CELEBES, AND BACK TO COMORO

and belong partly to the French and partly to the British. They are rugged and mountainous and heavily wooded. The land looks fine from out here. A small river is rushing down the mountain side. How I would like to go on shore there!

Monday, June 27, 1881

We have had no bread but hard-tack for the last three or four days as our baking powder is all gone and we can't get anything like that around here. We have plenty of flour but no yeast or baking powder. Mr. Gifford said, "Bob, can you make bread without yeast? I'd give a barrel of flour if you could make some like you made at Fayal." I said, "Give me about forty-eight hours and say nothing to the captain." I got a clean bucketful of flour, a quart of boiling water, some salt and made a batter. I left this in the bucket, covered it up with a cloth and left it on the shelf in a warm place to ferment. Mr. Gifford asked if the bread would be sour. I told him that anything that will ferment will make bread, even beans and corn. I learned that from my father when I was thirteen years old. He was a past master at baking.

Wednesday, June 29, 1881

I looked at my sour dough and found it fermenting fine, so I mixed enough flour with salt in it to make a stiff dough and set it to rise until morning. It looks all right, lively and springy, and the warm weather will help the dough.

Thursday, June 30, 1881

My dough was in good shape this morning, so I got a handful of sugar, a piece of butter the size of a walnut and mixed it in the dough. Made one small and one large pan of biscuits and a panful of good-sized loaves of bread. I put them all to rise and told the cook when to put them in the oven, not to open the door of the oven for fifteen minutes and be careful then not to get them too brown. The biscuits were ready for dinner and light as a feather. The mate came up from dinner, all smiles, and wanted to know if he could have bread like that all the time. The captain came down from the quarterdeck and asked me to show the steward how to make bread like that. I told the steward to save a good-sized piece of dough each time and put it aside to ferment. I told Mr. Gifford that my father was a fine baker and that he had shown me two or three ways to make bread without yeast. I said that I had worked alongside of him when I was ten years old and until I was seventeen, at which time we had a quarrel and I left home at his command.

At noon we sighted Bird Island and hauled aback on Mahé Banks in thirteen fathoms of water. The water was so smooth and clear that you could see the bottom quite plainly. Some of the fish are very pretty, all colors, yellow, blue, black, green and gold. They are of all shapes and sizes, and you could bear to watch them all day fleeting in and out among the coral branches. Large oyster shells as big as wheelbarrows could also be seen.

The captain lowered the waist boat and went on shore at Bird Island. We gathered three bushels of newly laid birds' eggs. I marked off a place and cleared away all of the old eggs, but in two hours the cleared space was covered with freshly laid eggs, which we picked up to take back to the ship. The birds were so plentiful that you had to push them to one side as they tried to alight on your head and shoulders. There is not a drop of fresh water on this island. There are a few coconut palms, but hardly any shrubs.

Friday, July 1, 1881

We ran along the Mahé Banks and sighted the Seychelles Islands which belong to the British. We hoped to get a run ashore but the captain said nothing and steering a course to the eastward, passed to the southward of them.

Today we sighted two whaling barques, one of them the *Osceola* of New Bedford. She has been out for three years and has had miserable luck. While cruising around the Mahé Banks, she had all of her boats smashed. We gammed this little barque. They seem like a good lot of men. The officers are mostly young except the mate who is a hard-looking old man with an ugly

JAVA, CELEBES, AND BACK TO COMORO

scar across his face, but Mr. Gifford says he is a fine man and a good whaleman. They have eight hundred barrels of oil on board and are going to the Sulu Sea where they have heard that whales are very plentiful. They heard that the *Minnesota* filled up there in a year, no longer than two years ago, although it is a great place for typhoons and hurricanes.

I took a sight for Mr. Gifford and got a chance to work out the observation while he was talking to the other mate. I was glad of the chance and made it Lat. 6° 41′ S., Long. 56° 27′ E., just south of the Seychelles Islands. When the captain returned to the *Kathleen*, he clapped on all sail and set the course to the eastward following the *Osceola*.

Thursday, July 7, 1881

We are still heading eastward across the Indian Ocean with all sail set. For the past few days I have been working on my checkerboard, using pieces of rosewood, ebony, satinwood, and sandalwood. Today I am working on another coconut dipper. It is still green so I carved a border of the walls of Troy. I am going to make the handle of ebony or rosewood tipped with ivory. The first dipper that I made has hardened and has taken a fine polish, like glass. Mr. Young asked if I was getting ready for housekeeping.

I replied, "I guess not, as I have no girl."

Sam Hazzard said, "Bob, you don't know what might happen."

Mr. Young chimed in, "Bob, Sam will speak a good word for you."

Sam said, "Let him do his own bargaining."

Mr. Young said, "Sam, we will all be up to your wedding whether we are asked or not."

Friday, July 8, 1881

I overhauled my boat gear today, whetting and oiling the lances and irons, greasing the rowlocks and sharpening the knives and hatchets. At dog watch, all four of us harpooners and the cooper went down into the steerage. Tommy Wilson started to sing some Yankee whaling songs. Then we all joined in together and sang "Annie Laurie," "Old Black Joe," "Maggie May," and wound up with "Auld Lang Syne." It made us all feel happy. When we came on deck, Mr. Reed said, "You fellows have been having a good time." Tommy said, "We know how to drive dull care away and be happy."

Sunday, July 10, 1881

Carrying all sail, it is all we can do to keep up with the little barque *Osceola*. She is a right smart sailer in light winds. It is a wonder that we did not see her earlier, as she was down around the Comoro Islands about the same time that we were.

The mate told the steward to make some coconut pie. The steward wanted to know if you cut up the coconuts like apples for apple pie. I had to tell him how to grate the coconut fine and mix it in a custard of eggs, condensed milk, and sugar. The steward wanted to put a crust on top but I told him that coconut pie should never have a top crust. Mr. Reed teased the steward about the Portuguese pie he once made and which had to be thrown to the pigs. Mr. Young said that the pies that I made did not go to the pigs, to which Mr. Gifford remarked, "At least not to the four-legged ones."

Tuesday, July 12, 1881

We are near the equator and it is pretty hot, but we are going along at a good clip night and day. Only when near islands at night do we shorten sail for caution. The other whaler, *Osceola*, is going right along with us, so there must be some agreement between the two captains, for we are both heading for some place east. We passed Christmas Island today. There are lots of birds in the air and lots of life in the water but not a spout of any kind. We keep four men at masthead from daylight to darkness. There is a bounty of five dollars for the man who raises sperm whales. Mr. Reed says that we are heading for the Sulu or Java Seas where it will be lots of fun handling sail, as it is squally most of the time.

JAVA, CELEBES, AND BACK TO COMORO

Thursday, July 14, 1881

The weather was very hot and as we approached the land, rather high land, the wind which was strong, became light. The *Osceola* was about a mile ahead of us and gaining in the light wind. I worked on my checkerboard all forenoon except for two hours when I was at masthead.

At one o'clock it was reported that the *Osceola* was entering Lombok Straits. When the *Osceola* was opposite Bandit Island, our man at masthead reported her in trouble as the flag at her peak was upside down and that she had no wind.

The captain took his spyglass, turned to the mate and said, "Canoes are swarming around the *Osceola*. We had better go to her quick."

The mate rushed the order to get the signal gun out of the lazaret because Malays in sixty or seventy large canoes had attacked the *Osceola*. As our ship drew nearer, we could see the canoes, filled with big brown men, armed with spears and knives. The lookout at masthead called out that if the wind held out we would reach the *Osceola* in ten minutes.

The mate, captain, and cooper were stationed on our quarter, the fourth mate with his crew forward, and Mr. Young and Sam Hazzard in the waist with their crews. Mr. Gifford told me to fill the try pots with water, fire up under them, get the water boiling and make haste, using my boat's crew to help. He gave me two muskets and told me to get lances out of the boats. I was then to get the deck pump, a small force pump, and have two men man it when the water was boiling.

He said, "Bob, don't let any of those natives on board. Shoot to kill. Scald as many as you can."

There was a good gang forward with Mr. Reed and Mr. King, all armed with muskets. The captain, cooper, steward, and cook were all aft. Frank Gomez got busy firing whale scraps, and the pots were soon boiling. Just as we passed Bandit Island at the mouth of Lombok Straits, ten large canoes dashed toward us. With the small cannon, the captain sank one alongside the ship. Mr. Young shattered another with a whale bomb gun and killed about six Dyaks or Malays. Some of the natives tried to climb up the side of the ship, but as fast as they clutched the rail, the cooper chopped off their hands. The steward shot one with a musket against his breast. Two large canoes came alongside where I was stationed by the fore rigging and try works.

I sang out, "Pump, Fernand, pump hard," and squirted the scalding water back and forth on their bare backs. I had to use a piece of canvas to hold the nozzle of the hose, it was that hot. The two canoes were emptied mighty quick. The natives dove and jumped over into the sea with screeches and yells that did not seem human. Three canoes cut under the bow and Mr. Reed lanced the men, one at a time. Mr. King shot them as fast as they stuck their heads above the rail. Peter was busy loading muskets for Mr. King.

We finally got down to the *Osceola*. She was hard pushed but we soon had the Malays going. The whale bomb gun and the little signal cannon sank about twenty-five canoes. What Malays were left started to paddle for shore as fast as they could. I tell you, they had a dread of the scalding water, for they were all naked except for a waist cloth. When I squirted the boiling water on their backs, it peeled off the skin and they jumped overboard with unearthly yells. These Malays are fine, big, strong, powerful looking men, tall and broad shouldered.

The *Osceola* was laying aback and I suppose waiting for us to come over. After I went to masthead, the captain lowered the starboard boat. It began to leak. One of the natives had run a spear through it. It had to be hoisted and the cooper had a job. The officers lowered the larboard boat and went gamming on board the *Osceola* to find out how she fared after the attack. They had four men killed, and half of the crew and all of the officers were wounded. Mr. Reed said when he came back that it looked like a hospital ship. Six of our men on the *Kathleen* were wounded, not dangerously, mostly on the arms. Mr. King, Tommy Wilson, and Ben Butler were among those hurt. Tonight we got through the Straits and out into the Java Sea. We got well off shore, shortened sail, cleared up the decks, and kept a sharp lookout all night.

JAVA, CELEBES, AND BACK TO COMORO

Friday, July 15, 1881

We are cruising toward the Flores Sea. The *Osceola* is about three miles off. In the morning the wind blows from the land and in the afternoon from the sea, but it is never very strong unless we get a squall, and then it comes with very little warning. We ran into one sudden quick squall and had to call all hands to move lively as we did not want to lose any spars. Land is in sight almost all of the time. As we enter the open sea, there is more room but the heat is intense.

We went on shore on a small island and got some yams, chickens, dates, fruits, and nuts. We also got some melons and canteloupes, all large and sweet and of a very fine flavor. One bolt of loud calico paid for all.

I was dozing in my bunk when I heard the cry from masthead, "There she blows, a school of small whales!" We lowered away four boats and hoisted sails, but the wind was so light that the boys had to get busy and paddle for all they were worth. The three boats from the *Osceola* were coming on to the school of whales. The whales went down out of sight just before we got to them. After the whales came up, the seven boats scattered and each one got fast. The whales put up some grand fights with heads and tails. They rolled over on their backs and snapped their jaws, and the boats had a hard time to keep clear. Our fourth mate had his boat stove badly but he killed his whale. Our whale went into a flurry, then went down, coming up almost under our boat, and lay there. Mr. Reed killed it with a lance. Mr. Young seemed to have no trouble at all with his whale. The mate of the *Osceola* got two whales, but their captain had his boat stove badly. Mr. Young had to go to his aid and tow his whale alongside the *Osceola*. These whales were small. While cutting in, I never saw so many sharks, swarms of them all around the whales, biting and tearing the carcasses after we had peeled off the blubber.

Saturday, July 16, 1881

It is very hot weather to be firing up and boiling oil. All hands are sweating, although most of the men are wearing nothing but knee-length dungarees. Some natives came off in two canoes to trade but the captain would not let them come on board and drove them off. These natives are treacherous and not to be trusted. They are not negroes, for they are coffee colored and have long straight black hair.

MACASSAR, CELEBES

Sunday, July 17, 1881

After breakfast we stowed down forty-eight barrels of oil. After a general cleanup, we gammed the *Osceola*. She is a nice little barque and handy. She stowed down sixty-two barrels. We passed a small low-lying island with lots of trees, bushes, and a scattering of lofty palms. Birds in thousands were swarming all around, diving and catching fish in the beautiful lagoon. When cruising in this heat, nobody cares to do any more work than actually has to be done. I took my checkerboard on deck to work on it, but only worked ten minutes and took it below again. I did not feel like working. The sweat kept dropping off the end of my nose until it wet the deck.

We passed quite a large barque today and had quite an argument as to whether it was British or Yankee. I said, "If you'll look with the glasses you will see something that speaks plainer than the flag, a round Plimsoll mark that is on all British ships." It proved to be the *Kingston* of Bristol. This morning we had a bad squall. It only lasted half an hour but was very severe, and the men had to be called down from masthead. When they eventually went aloft again, they reported the *Osceola* in trouble, and that she had lost some of her spars. We ran over to her and found that the mizzen mast had been carried away, smashing the starboard boat. Mr. Gifford took fifteen men over to help clear the wreck. They saved all the rigging but the lanyards were all parted. The topmast was all right. He helped get her to a place called Macassar on the island of Celebes, where they hoped to get a new mizzen mast. When we got to Macassar no one could be found who could do the job, so the *Osceola* will have to go to Singapore, cruising along the Java Sea to the westward. We will go along part of the way and leave this place in

JAVA, CELEBES, AND BACK TO COMORO

the morning. There are many islands, making it a dangerous place in the night as there are no lighthouses to go by. You have to keep a very close watch on the compass.

Thursday, July 21, 1881

We parted company this morning with the *Osceola*, she steering north and by west for Singapore, that holy terror of a place which all sailors like, and which is busy with ships from all parts of the globe. We steered south and by west at right angles to the *Osceola* through Sunda Straits near Batavia. We saw high mountains as we passed along. In the afternoon we were out in the open water again where the heat was not so oppressive.

ALDABRA ISLANDS

Monday, August 1, 1881

We sighted the Aldabra Islands in Lat. 9° 23' S., Long. 45° 50' E. and went ashore. We met an old Frenchman and his family. He has a nice little house, lots of chickens and several pigs which he feeds on coconuts and corn. He said that sometimes he does not see a ship in a year. We got two small black pigs, twenty-five dozen eggs, and thirty chickens. When the captain wanted to pay him, he did not want to take money because he could not spend it. When asked what he wanted, he replied, "I would like some cloth for my wife to make a dress, and some rope for my boat." We took him out to the ship and the captain gave him a suit of light clothes, twenty-five yards of calico, and a piece of whale line twenty fathoms long. The Frenchman said it was too much, so when we rowed him ashore, he would not let us go until he gave us some oranges, coconuts, five large melons, and a lot of yams. He stood on the beach, uncovered, until the boat reached the ship.

AT JOHANNA, COMORO ISLANDS, ZANZIBAR, AND BACK TO JOHANNA

Thieving natives—Slave stealers—English gunboats and slavers—Thrown out of mosque—Rafting water—Shark's stomach—Painting ship—English sailors—Visit to mosque—Sugar plantation—Slaves cruelly treated—Dinner at palace—Collection of arms—Old Arab nearly stabbed—Rescue and fight—Rewarded by old Arab—Zanzibar—American Consul—Slave market—Slave dealer killed—Escape of white slave—Coast fever—Hurt by windlass—Large whale—Yarn of pirates and slavers—Arab insulted and fight with knives.

CHAPTER V

AT JOHANNA, COMORO ISLANDS, ZANZIBAR, AND BACK TO JOHANNA

Tuesday, August 2, 1881
We reached Johanna in the Comoro Islands early today. It is a beautiful island with high mountains and a few scattered houses here and there along the coast. The trees grow down to the water's edge except in a few places where there are steep rocks. You could smell the odor of spices from the land. We got a range of chain out, shackled it to the port anchor and let go in four fathoms of water. You could see the bottom quite plainly, the water was so clear. When the rigging was all clear, we harpooners got orders to take all small articles out of the boats, such as knives, hatchets, rowlocks, or anything that could be lifted, and to hoist the boats on the upper cranes. The natives here are a bunch of thieves and will steal anything they can get a grip on. The people here are nearly all Mohammedans, mostly Arabs. They wear a long dress, a tunic they call it, like a nightshirt, spotless white. They can lift anything with their toes. Sometimes they wear sandals which are only soles with no straps and with a button like a spool between the big toe and the next which holds the sandal to the foot. They all wear turbans on their heads, some of which are enormous. A lot of dugouts, with mostly Arabs in them, came along after breakfast. These men are dark, almost black, but they have sharp features with large hooked noses, not broad like the negroes. Only these Mohammedans still hold slaves who live back away from the coast. The British gunboats are after them all the time on the sea but cannot interfere with them on land. The British make quick work of the slave stealers, for they hold no mercy for them. They send sub-lieutenants in launches into the coves along the coast wherever they think a dhow would try to land a lot of negroes. These launches usually carry a small brass gun on the bow and are manned by a crew of ten men and a coxswain, and sometimes remain away from the gunboat for several weeks at a time. They dart on a dhow full of slaves stealing in

to shore at night. As the dhows may carry a crew of anywhere from fifteen to twenty men, and the very large ones thirty men, it is readily seen that the British have no easy task.

One of the men from one of these gunboats told me about his young sub-lieutenant who chased one dhow into the mouth of the river and found another dhow there lying in wait for them. The sub asked his men if they wanted to have a "go" at that bunch or get out as it might be their end. "If you say fight, fight it is to a man," he said. The men replied, "We are British." They had orders to keep both dhows on one side so they would not be surrounded. They had a hot fight for an hour, but it seemed like three. Two of their men were killed. The young sub was wounded in the arm, and the rest of the men were all crippled. Only five men of the slaver's crew were living, and they were tied together and shot like dogs. They found over one hundred slaves chained in the two dhows. They were freed of their chains and placed on one dhow while the dead were put on the other dhow and burned. The slaves helped to do the work. At daylight they hoisted sail on the dhow, and in the afternoon met the gunboat. Today, the young sub is in command of the *Sea Gull*. The men got twenty pounds each as their share of the money found on the dhows.

Wednesday, August 3, 1881

Our deck is crowded with Arabs and the men forward are beginning to miss things. These people seem to steal for the love of stealing. I went on shore to see the old town which is said to be four or five thousand years old. The streets are narrow. The houses have thick solid walls of stone and flat roofs, but no matter how hot it is, the insides of these houses are cool and comfortable. Some are furnished grand, others scant, maybe having only a lounge and a few rugs or mats on the floor.

I went in to a fine-looking mosque and did not get far when I was yanked out by big natives who were gabbling away at the tops of their voices, but when outside they let me go. I met a Mohammedan talking in English to Mr. Young, and when I told him about it he looked solemn and then laughed. He told me I should have taken my shoes off and washed my feet before entering the holy place. He said that I should leave my shoes outside and that no one would touch them, then go down the marble steps to the water where I could wash my feet. I did not go back today but will have another try at it sometime.

Instead, I took a walk with Frank Gomez. We went down a beautiful valley with a small stream of pure, cool water coming down from the mountain top. Birds were singing in the trees. They were quite tame and had such brilliant colors. There were small kingfishers, oh so dainty, and parrots that mocked you. We got some fruit and sat down to rest at the stream and appease our hunger. Some of the trees were very beautiful with bright green shining leaves and clusters of red berries. We found that these were clove trees by the round balls on the end of the cloves, now bright red but black when ripe. Oh, how nice was the fragrance of these and other trees close by! As we came down near the shore, we saw a tree full of large flying foxes, hanging to the branches, heads downward with their wings spread out. There were hundreds of them in among these large trees. The flying fox looks like a big bat but has a head like a small black-and-tan dog with the ears standing straight up as if listening.

Thursday, August 4, 1881

The weather stays pleasant while we are at anchor here. I noticed an English ship in the cove about two miles from here, loading sugar. There is no wharf and they have to load her by lighters as the water is shoal. On our deck, in open casks, are oranges, sugar cane, pineapples, coconuts, peanuts, bananas, lemons, tamarinds, mummy apples, guavas, to all of which our men can help themselves. There are bunches of bananas hanging on the rigging, and piles of dates and figs for those who want them.

The captain and mate are on shore all the time, living up on the mountains. The rest of the officers are hardly ever on board. Where they stay I do not know, as there are no places of amusement. Sam Hazzard, Tommy Wilson, and I are on board every night.

The cooper got some casks becketed to raft fresh water to the ship. We towed them on shore to a small river, filled them and got them back to the ship this afternoon. The casks were eight-barrel casks and we got about a hundred barrels from the river only a quarter of a mile away.

Sam caught a very large shark, a monster, which took all hands to hoist on deck. We killed and opened it. It had everything inside it but, as John Brown said, an elephant. There were old boots, tin cans, a piece of chain, three or four whole fish, a sheath knife, a silver fork, several buttons, and a lot of fish bone. I gave John Brown the silver fork, and the sheath knife, a right good one, to Frank. Sam chopped the tail off and nailed it to the end of the flying jib boom. When we threw the carcass overboard, there was a scramble among the natives to get a hold of it, and who tugged it ashore with their dugouts.

Friday, August 5, 1881

The men were busy today painting the ship inside a dark cream, the waterways, hatch combings, bitts, and pumps a dark blue. The captain came on board and called me below to ask if I would paint his cabin. I asked him how he would like it done. He told me to use my own judgment, so we moved his things out of the cabin and I washed off the old paint good. Then I mixed and strained my paint and washed and cleaned all the brushes, ready to go to work tomorrow.

Thursday, August 11, 1881

I gave the after cabin a coat of light cream and the panels a light lilac. I told Mr. Young I wanted some gold leaf to trim the narrow bead between the panel and the stiles. He said, "If there is any in this place, I'll get it for you." He found it, so I striped the bead around the panels with the yolk of an egg and laid the gold leaf on it. I cut a stencil of fleur-de-lys, which I painted in pea green, at the top and bottom of each panel. The floor got a coat of terra cotta, and when it was dry I lined it in thin lines of dark blue to give the effect of tiles. Mr. Young said that the work was too good for an old greasy whaler.

After dusk, a boat hailed us in English, "Barque ahoy," and came alongside. There was the lieutenant of the British gunboat *Sea Gull*. He came on board and told us that he was looking for a large dhow that was supposed to come in somewhere near here tonight or tomorrow night. We got some coffee and something to eat for him and his twelve men. We gave them some fried fish and some fried sweet potatoes and they thought it was a treat. They liked our hard-tack so we gave them a bagful to take with them. Then we had some singing. Two of the Englishmen were splendid singers. Tommy Wilson gave us a song and he was followed by the lieutenant who had a grand voice, after which Sam Hazzard sang a whaler's song, "Come All Ye Bold Seamen."

Saturday, August 13, 1881

After breakfast, Tom, Sam, and I went ashore to look around the place. When we came to the mosque, we took off our shoes, washed our feet and went in. There were many fine rugs and carpets on the stone floor. Here and there were a few people kneeling at prayer. I looked around but saw no altar or pulpit, nothing but fine rugs and bare walls, all very clean even if it was said to be thirty-five hundred years old. I knelt down on a rug and said a prayer. When I got up, one of the Arabs came and asked if I was a Mussulman, but I shook my head.

We walked up the country road five or six miles to a sugar plantation where we found two white men. One, a doctor, was a Yankee from Connecticut, Nelson by name; and the other, his partner, a Scotchman. They had planned and built the mill with the aid of black slaves. It's a fine plant in a beautiful valley. The two men live in a long, low stone house, one story high, with a fine garden with vegetables of all kinds, that grow all the year round, tended by an old slave whose wife cooks for them and keeps the house neat and clean. At the side of the house is cold water bubbling out of a spring through small white pebbles.

They have a broad porch with comfortable cane chairs. The

two men showed us over the plantation. The sugar cane grows all the year round. It is cut, taken to the mill, and crushed between large steel rollers, coming out nearly dry. After spreading the waste (bagasse) on the fields to become thoroughly dry, it is used as the fuel to run the large, powerful engine. They let us look all over the place. We saw the sugar boiling and the slaves putting it into bags. Other slaves were making bags and mats to hold the sugar. There is a small river running alongside of the mill, where the sugar is loaded. All of the sugar goes to a firm in Greenock, Scotland. Every three months they load a ship and send it away. The engineer is clever, no doubt, but a clever devil, for it's cruel the way he treats those black slaves. Some tried to run away and he put heavy shackles on their feet and hands, and three links from the wrist to the ankle. For food, they had a dish of water and a handful of raw rice.

When I saw four of these slaves naked and shackled, it made me so angry I could have killed the Scotchman. The feeling came over me, what right had these two white devils over these poor blacks to treat them like this when they had committed no crime. Any brute was treated better. These slaves are gentle and kind to the peaceable white man. We were to stop all night, but when I saw how these poor blacks were treated, I could not go and eat supper with those fiends. We went back to town and had to call the ship to send a boat for us. Mr. Young said he thought we were going to stay all night. When Sam told him what I thought about the slaves, he said I was too chicken-hearted. He also told me that the captain had been on board and was much pleased with the way his room was painted, and that he was coming on board tomorrow to take Sam and me for a big time or a dinner and that we were to shave and dress. If it was the sugar plantation where he was going to take us, I was going to tell him I was not going, and why.

The view up on the mountain was grand. The dates and coconut palms, the clove trees with their bright green leaves and clusters of red berries were beautiful. The smell of the ripe fruit, the orchids, and the cinnamon trees so fragrant and the sight of the bright-plumaged birds flitting from tree to tree was a pleasant memory, but tonight as I knelt down, I kept thinking of those poor slaves.

Sunday, August 14, 1881

The weather is fine and clear. Sam Hazzard and I up and had a shave and a good swim. We rubbed one another down well, then dressed and had an early breakfast. At eight o'clock the captain came on board and told me how pleased he was with his room and said it was far too grand. He told Sam and me that he wanted us to come with him to a feast at the King's palace.

Donkeys, in charge of a little black boy, were waiting for us at the foot of the hill near the mosque. We mounted and started up the hill along a little footpath, through a thick wood, across a beautiful valley with large trees, dates, figs, and tamarinds. We passed through a small village of native huts and about a mile farther on came to a large town with a high wall of stone. We went along the wall quite a distance until we came to a large brass double gate with a small one on each side. There was a guard of Arab soldiers at the gate as we went in. They saluted by raising their scimitars at "present arms," which we acknowledged with a salute of the hand. We were led up a walk lined with rose bushes to a fine stone house with large marble columns along the front, and broad steps leading to a large door studded with silver. A major-domo met us at the door and led us, without saying a word, into a large hall or room. There sat the King on a raised platform. He was a fine-looking old fellow with a long beard. Around him were about a dozen officers, all in gorgeous uniforms, green and gold, blue and yellow. The King spoke good English and had quite a talk with our captain.

Sam and I looked over the palace. It certainly was fine, more magnificent than any place Sam and I had ever laid eyes on. The walls were beautifully decorated and had many fine paintings hanging on them. In one large room, like a museum, there were guns of all descriptions; pistols, large and small; daggers, long and short, crooked, slender or broad; spears of all kinds; numerous swords, long and slender, broad and curved; the basket-hilted sword of the Highlands; claymores; broad scimitars of the Turk, some of the hilts of which dazzled with gems; and shields including the round targe of the MacAlpin of brass-studded bull's hide. No two shields were of the same shape and there must have been fifty or more, some of brass and steel, others of rawhide. There were spears by the hundreds, some having queer-shaped heads. There were coats of armor and chain mail, helmets and battle-axes by the hundreds, some with long handles like the English bill-hooks, others were short battle-axes. There were headpieces of iron and brass with long plumes of horsehair. From blunderbusses with mouths like bugles, and flintlocks with long or short barrels and inlaid stocks, there were all kinds of guns down to the present-day English Martini-Henrys, and repeaters.

Finally dinner was announced. It was served to about fifty people in a large banquet hall. The service was all of sterling silver, probably cast silver. I noticed that it was made in London. It was a good feast although half of the time I did not know what I was eating. No wine or liquor of any kind was served, but there was plenty of cocoa, coffee, a very sweet lemonade, and a drink made from tamarinds that was sour but very pleasant.

The captain said the King would like to speak to me. It seems that the captain told him that I was a Scotchman. As the King had received his education in Edinburgh, he liked to meet any person that he could talk to about Scotland. When I told him that I came from the Highlands, he said that he had gone to the Highlands with a young man who was at the University with him. He said it was a fine country but very cold. I asked him in what part of the Highlands he had been. He said that they first went to Inverness by train and then by coach through rugged mountains and a large forest, called the Black Forest, to a small but very beautiful place called Lochinvar. His friend's father lived away back there among the rocks in a large castle. The King shook hands and said that it was a pleasure to have talked with me, but as for going back to the ship that night, he would not hear of it. The captain did not need much coaxing to stop until morning and said that we would start back after breakfast.

Sam Hazzard and I took a walk through the grounds around the palace. They were beautifully kept and in good taste. I never smelled anything like the fragrance of the roses, fine large ones of all colors and so many of them.

Monday, August 15, 1881

After a breakfast mostly of fruit and coffee, we found the donkeys waiting for us at the door. The King bade us goodbye. He was a fine old man sixty or seventy years old, but straight as an arrow and his skin not too dark from the tropical sun. When we got to the gate, four Arab soldiers with long spears, on fine horses, escorted us, two on each side. As they could not speak English, we had no talk with them. It was a pleasant ride trudging along on the donkeys, watching those fine, big, spirited Arabian horses prancing at our side. After taking us through the town, the soldiers turned back and we continued on the donkeys to Johanna. The captain asked how we liked the ride and we told him it was a treat. He told us that he had not wanted to go alone.

Tonight when the captain went ashore, he took me along with him. When near the mosque, we heard a muffled and smothered cry, then a shriek. Close by, in the shadow of the wall, two men were fighting with an old man who was trying to cover his head. I walked up and gave one a kick in the stomach. It rolled him over and I gave him a kick in the ribs as hard as I could. When I turned around, the other fellow, with a knife about a foot long, was trying to stab the old man. I took off my coat, wrapped it around my left arm and went at him. He made a stab at me. It stuck in my coat. I let him have a smashing blow in the jaw as he lunged at me. He went down like an ox and dropped the knife. I picked it up and gave it to the old man. It belonged to him but the fellow had snatched it. The captain said he could do nothing, for it had all happened so quick. He thought I was quite handy, but one has to work quick in cases of that kind.

Tuesday, August 16, 1881

The weather is fine and the wind light. After breakfast, when all of the officers were on board, everyone hustled to get the running rigging down on deck, the gaskets off the sails, heave the anchor and hoist the fore and main topsails. The men at the windlass gave a chantey and got the anchor on the bow.

An old man came alongside in a canoe and climbed on board looking all around for me. All excited, he ran up to me and salaamed and salaamed. He pressed something in my hand and patted me on the shoulder. The ship was under way, nevertheless he went over the side into his dugout and paddled ashore. When I opened the package and looked at the contents, my eyes dazzled. There was a very small cup, very fine, that would hold about three teaspoonfuls. In it were two diamonds, two pearls, and two or three red stones. I went below and put them in the till of my chest and said nothing to anybody.

After clearing up the deck, lookouts were placed at fore and main mastheads and we stood out to sea. A cask of oranges was lashed on deck, besides baskets of peanuts, pineapples, and all kinds of fruit for those who wanted to help themselves.

ZANZIBAR

Wednesday, August 24, 1881

This morning as we headed in toward Zanzibar, I saw the British gunboat *Sea Gull* going in towing two dhows. We hauled aback at the mouth of the harbor. The captain told me to come along with him. We lowered the starboard boat and had a pull of about four miles in the broiling hot sun to the landing stage. After telling the men not to leave the boat as the smallpox was very bad, the captain and I went to see the American consul. The streets were narrow and filthy dirty in this old city. Some of the Arabs threw stones at us as we passed up the street. After talking to the consul for a while, the captain told me to go back and take the men and the boat to the ship but to come in again tomorrow morning, but the consul persuaded the captain to let me stay all night and said that he would send a man down to the boat to tell them. I was glad, for I did not like the idea of going down that narrow street alone and having the Arabs sling stones at me.

After lunch, the captain suggested that we go to the slave market. I had only my shirt and pants on and protested that I was not dressed up. He said, "You will see some with much less clothes on than you have." We came to a high-walled enclosure and passed through a wooden gate into a large hall with dirty benches around the walls. A platform about three feet high with steps leading up to it stood in the middle of the room. We hardly got seated when an official stepped up on the platform and made some announcement. From a side room, an Arab came leading four large, strong, black men, all naked. He led them up to the platform where they were auctioned off. Following more men, a big, strong, black woman with a baby in her arms and a little boy about three years old, were sold. The final sale was a woman, almost white, a fine-looking, well-built woman. An old Arab with ugly yellow teeth and long fingers went over and looked her all over and laid his hands on her bare breasts. When he did this, she hit him a slap in the face that sounded through the hall. He drew a whip and was going to thrash her but a man thrust him back on the bench. However, the old Arab bid her in and bought her. We went back to the consul's house, but very little sleep did I get that night for thinking of that black mother and her two babies, and the near-white woman that the old Arab had bought.

Thursday, August 25, 1881

As I lay in bed this morning at the consul's, I heard a terrible hubbub and an awful jabbering of tongues. The consul said that there was something unusual going on, so he called his little black boy, questioned him and then sent him out to find out what it was all about. In about twenty minutes he came back with a horrible tale. It seems that the old Arab who bought the white woman was found in bed with his throat bitten through and lying in a pool of blood. The near-white woman was not to be found anywhere, although they had searched the city for her.

Monday, September 5, 1881

For ten days we have been cruising around the Comoro Islands looking for whales. Although there is a bounty of five dollars for the man who sights a whale, we have had no luck. Mr. Gifford and Manuel King are both very sick with the coast fever. At one o'clock I went to masthead and sighted a whale almost at once. At the second rising, the larboard boat got fast and it sounded again. When it came up, the three boats were close and soon lanced it in a vital spot. It was a much larger whale than usual.

Tuesday, September 6, 1881

The weather is pleasant under the lee of Comoro Island. Called all hands at daybreak and started to cut in. We got the body blubber all in by ten o'clock. When we started to heave in the large head of this big whale, the walking beam on the windlass broke, hurting the cook, Frank, and breaking Peters' arm. I was hurt on the shoulder. The head was so heavy and the strain so great that everyone who could be spared was at the windlass heaving with all his might. When the timber broke it was lucky no more were hurt, for we were all crowded together. However, we finally got the head in and started the try works. I set Peters' arm, put on splints, and bandaged it.

Friday, September 9, 1881

What with the hot weather and the hot oil, it kept the cooper busy driving the hoops down on the casks, for they shrink badly under the heat. I called all hands to stow oil in the after hold. It was a hot, greasy job but we stowed away 105 barrels of oil from this huge whale. After we cleaned up the deck, I had to go below on account of my jaw being all swelled up like a bucket and aching awful.

ZANZIBAR, AND BACK TO JOHANNA

Tuesday, September 13, 1881

Yesterday we got another whale but I could not go because I got so bad with my head. Today the abscess in my jaw broke outside and inside, and I feel a great deal better. What a relief to have the pain all gone!

Friday, September 16, 1881

From masthead, in the afternoon, I watched two dhows close in shore come together fighting. One dhow crashed into the other, and with the glasses I could see the men knocking one another overboard. They had a fierce time of it. One of the dhows sunk level with the water.

Johanna is on the weather quarter and Mayotta on the beam.

JOHANNA

Monday, September 19, 1881

Lying here at Johanna and having nothing much to do, we lowered a boat and went sailing down the coast. The scenery was grand. We ran in at the mouth of a small river that came running down from the north. The water was cool and clear and we all had a good drink. We walked a little ways and picked some oranges and bananas. Some of the coconut trees being slanted were easy to climb, and we gathered a lot of coconuts.

Peters went on shore today to have the doctor fix his arm. He told me that the doctor said I had done a good job and that it was mending fine.

Friday, September 23, 1881

All hands were busy getting a raft of water casks from the shore and getting them stowed down below. Another gang got wood and stowed it away. Sam Hazzard and I, during the latter part of the day, went fishing and killed four sharks. We sold them to the Arabs for five rupees and they thought they had a bargain. We divided the spoils and had a supper of fresh fish, but not shark. We have not come to that yet.

Sunday, September 25, 1881

Got half of a native humpback cow today and had steak smothered in onions. It ate fine with roasted sweet potatoes. Also got some guavas and some taro. I wrote a letter home today. That makes five I have sent from here. Sam has written about ten to Massachusetts. Sam and I had to look after the ship, for nearly everybody else had gone on shore. I cooked dinner and gave the men forward the same as we had.

Wednesday, September 28, 1881

The British gunboat *Sea Gull* anchored here today but went away at three o'clock leaving a launch with ten men here alongside of us. We asked them to come on board. The sub-lieutenant was a young man who told us they were after pirates and slavers. He told us of one they were after about eighteen months ago. This fellow was a regular devil and very cruel. He had a schooner rigged like a dhow and very fast. It was fitted up fine, having two brass guns on her and carrying a crew of twenty men. When they boarded this boat for the first time, she looked like a coaster and had on board only four men and an old man with a long white beard. The cargo was all in plain view up to the hatch combings. Later, from one of the men who had been captured when this rascal had attacked a small brig, they got a good description of him and his boat. They caught him one morning standing out from Zanzibar. When they boarded her, there was the same old man with the long white whiskers. With the glasses, they had counted fifteen men on deck, but when they boarded her, there were only five to be seen. The sub-lieutenant made as if to shake hands with the old man, but instead quickly snatched the whiskers from his face. With a dagger in his hand, the old man made a stab at him but lost his hand at the wrist as the officer made one clean cut with his sword. The old man let out a yell and his crew came tumbling up from below armed to the teeth, but the sub-lieutenant's men were at the hatch and ready for them. In fifteen minutes they had the best of the crew, mostly white men, only seven Arabs. The hold was full of negroes, hand-cuffed to chains fastened to ringbolts in the timbers. The poor blacks were sick, but after getting them out on deck the fresh air soon restored them. The whole crew of the slaver was shot and their captain was made to walk the plank. He walked over into the sea with a laugh. The negroes were taken back to Africa, where they came from. An interpreter told how one of their own chiefs had sold them to a Portuguese slave-trader. With a force of thirty bluejackets, the sub-lieutenant marched to the chief's place, got the chief and, after a fair trial, told those who had been sold as slaves to hang him.

The sub-lieutenant and the bluejackets went back to the coast, and on examination found that the boat they had captured had two brass guns in the hold, together with arms in abundance, swords, powder, all kinds of clothing, bales of silk, calico, and cloth. There was a large amount of money in gold and silver, one box full of rupees and a small box full of bags of gold. Just how much I don't know, for the officers got the largest share. The sub-lieutenant and his men each got one hundred pounds sterling.

Sam Hazzard then told about the time when the Arctic whaling fleet of thirty-two ships was crushed and lost off Blossom Shoals. This happened when he went up as harpooner on the *Cleon* and came down on the *Helen Marr*. Sam said that about the thirtieth of June, 1871, the Arctic whaling fleet passed through Bering Strait following the whales north. The ships neared Icy Cape in July, but the ice prevented them from reaching Blossom Shoals. Later the ice opened up and permitted most of the fleet to approach Wainwright Inlet. The wind shifted and the ice nearly pinched the fleet. On September 2nd, terrible gales drove the ice in shore, closing down on them, crushing thirty-two vessels. Not a life was lost out of the 1,220 men, women, and children. Yet to save them it was necessary to drag the boats over the rough ice packs or row through the open strips of water some ninety miles to reach the unscathed remainder of the fleet located near Blossom Shoals. To house the survivors on the remaining vessels meant that they were so crowded that many had to sleep on the open deck above a blizzard-swept, icy sea. New Bedford's loss in this disaster was over a million dollars.

Then Tommy Wilson started to sing "Rolling Home to Merry England." The sub-lieutenant followed with "An English Sailor's Song," and Sam Hazzard sang a whaling song, "Rolling Down to Old Mahé." After that everyone sang "Annie Laurie." The sub-lieutenant called one of his men to come into the cabin and sing a song that he composed on board the *Sea Gull* about the pirates in the tale just told.

> His voice was low, his smile was sweet
> He had a girl's blue eye.
> Yet I would rather meet
> The storm in yonder sky,
> For there was blood on his right hand
> And in his heart was death.
> He knows he rides above the grave,
> Yet careless is his eye.
> He looks with scorn upon the wave,
> With scorn upon the sky.
> Great God! the sights that I have seen
> When far upon the main.
> I'd rather that my death had been,
> Than see those sights again.
> Shrieks have risen from the hold
> When human aid was none.
> Pale faces glimmer in the sun
> As giant waves sweep on
> And sink to rise no more.

It was one o'clock before we turned in to sleep and dream of pirates. I seldom dream at all but I dreamed all night that I was up in Hudson Bay cruising among the ice floes, or back in Philadelphia.

Saturday, October 1, 1881

We are still at anchor off Johanna. The fourth mate got a raft of water. It costs nothing so we got a good supply. The captain and the rest of the officers came on board. As usual, the deck was crowded with Mohammedans. I was standing amidship watching them to prevent their stealing anything, when I saw one of the Portuguese forward of the try works throw a piece of salt pork and hit one of the Arabs with it. The Arab looked around and saw no one but me. I was laughing. He thought it was I who hit him. To have spit in his face could not have been more of an insult to an Arab, for he would not touch salt pork by any means. He quickly drew a large knife and almost got me unawares. He made a cut at me and blood flew. I grasped my sheath knife and rushed at him and in trying to grasp his knife, cut some of my fingers. I cut his arm and got another jab in his shoulder. Mr. Gifford looked on, grabbed one of the whale guns, but the Arab dove overboard and made for the shore. Mr. Gifford was white as a ghost. Because of the blood all down the front of me, everybody thought I was badly hurt, but I was all right after the captain put a couple of stitches in the cut. The decks were cleared of these people, the anchor was hove short and we were soon off before the wind with all sail set.

FROM COMORO ISLANDS AND MOZAMBIQUE TO ST. HELENA

On shore at Europa Rocks—Wreck—Goats—Making trinkets—Scurvy on German brig—Thanksgiving day—Total eclipse of sun—Masthead punishment—Pipe overboard—On shore at St. Helena—Transferring oil—Visiting friends—Soldiers visit the ship—A welcome by Lady Ross—Dance at the barracks—Tipsy women—The Governor's garden—Trip to Napoleon's tomb—Another dance—Drunken bully—Rough and tumble fight—Lots of whaling ships in harbor—Gifts to Scotch friends.

CHAPTER VI

FROM COMORO ISLANDS AND MOZAMBIQUE TO ST. HELENA

Sunday, October 2, 1881
We are heading south between the islands of Comoro and Mohilla with the boats and the gear all in good shape and ready for whaling. Some of the men are making brooms of coconut fiber. These make splendid brooms for scrubbing the deck. Mr. Gifford caught a large dolphin that was following the ship, with a hook and line. The men seem to have had a right good time at Johanna. Some of them picked up some curiosities, others rare woods like sandalwood and rosewood.

Wednesday, October 12, 1881
For the past ten days the weather has been very pleasant, but we have seen no whales. The men have been busy on the sails and rigging. The cutting falls were overhauled. Chafing mats were put around the masts. We worked on the main topmast staysail; had the sewing on the pennants all chafed and rewormed; and served, parceled, bent on and tightened the stays. We made a coconut fiber rope for a cross-deck tackle for blubber, for no matter how greasy or oily it gets, it will not slip through your hands, and that's what we need badly when cutting in a whale in bad weather. We strengthened the foot of the jib and overhauled the foot ropes. New ropes and stirrups were put on the topsail yards. The fore topgallant sail was taken down, reseamed and patched. We took down the foretopmast stay and served it. The fore and main topsail lifts were taken down on deck, sewed, tarred, and hooked on again. We also strengthened and patched the fore topsail.

Friday, October 14, 1881
It's blowing hard today with a heavy sea running, a kind of cross sea, choppy and lumpy, that made like to pitch your guts out up in the rings. The ship tumbled about so much that you could not keep your footing without holding on to something. Some of the foremast hands are sick after the spell of smooth

water. Some of the hands are under the boats to keep dry, that is, under the spare boats that are upside down on the afterdeck where the men are handy for a call if needed.

EUROPA ROCKS

Friday, October 21, 1881

Yesterday it cleared up after a week of gales and rough weather. Today the man at masthead sighted Europa Rocks, so we stood in towards the island and hauled aback about a quarter of a mile from shore. The larboard and waist boats went on shore with their crews, for turtles. They brought off three large turtles, some fish, two goats, some mangoes and mummy apples. There was a wreck on the island that was not there when we went on shore going north. She must have run on either in a heavy storm or with all sail set, for she was well up on shore. She was stove in forward and pretty well stripped of everything movable. The deck fittings were broken and the hatches off. All of the yards were gone except the fore and main and they were hanging cockeyed. Even the topgallant masts were gone. Things down in the cabin were all broken open. The running rigging was gone. Both anchors were lashed inboard. She looks like a good sound barque and is marked fifteen hundred tons. The two goats that we got might have belonged on board of her, as they seem quite tame and appeared glad to see our men. We got a lot of eggs, good and fresh, and some large spider conch shells.

Saturday, October 22, 1881

At daylight, we set sail to the south. We gave the goats a good feed of hard-tack. We killed a turtle for dinner and got two buckets full of eggs from the turtles on deck. There are lots of birds all around. I like to watch them, especially the man-of-war birds, all black with a white ring around the neck. These birds are very swift and powerful and dart around like lightning.

Monday, October 24, 1881

Sighted a school of cowfish going along as if in a hurry to

get somewhere, jumping and playing. All at once they disappeared from sight. Not a one could be seen anywhere.

Tommy Wilson is making an ebony cane with an ivory head. Mr. Reed is making another of rosewood with an ebony head of a negro. He is also good at carving things out of ivory. He has made some very fine things, like a rosewood box inlaid with pearl shell and ivory, on the lid of which was a compass with all the points, the cardinal points being inlaid with tortoise shell. The lid slid off and on in a groove and locked with a small steel spring. Sam Hazzard is making a wheel jigger that Mr. Young says must be for his girl to help make her pies with. Peters is making a bone cane with an ivory handle. Three or four of the boys are making tortoise-shell watch chains. One fellow is making bracelets that Mr. Reed says are napkin rings and the cooper says are handcuffs.

Sunday, November 6, 1881

We have been having a lot of bad weather lately, but today it is pleasant. Sighted four ships today, all headed southwest, also several finbacks and a school of grampus. We are heading to the east, south of Madagascar, in Lat. 32° 55′ S. One of the men was fooling with one of the goats and teasing it when all of a sudden it turned and butted him and sent him sprawling. When he got up, the goat chased him and he had to go up the rigging. He called to some of the men to take the goat away but they just laughed. He must have done something mean to the goat. Ordinarily the goat is quiet and will eat hard-tack out of your hand.

Friday, November 11, 1881

At sunrise we sighted a French barque, a large one, as neat as a yacht, with clean bright spars, brasswork shining and glittering in the sun, and all of her officers in uniform. In the afternoon a German brig came along and spoke to us. She was bound for Natal from Calcutta. There were three men on board of her who were down with the scurvy as they had no fresh meat and no lime juice. We lowered a boat and took the goats and some lime juice to them. They wanted to pay for the stuff but Mr. Gifford would not hear tell of it. They were very thankful and threw a bag of fine coffee weighing about twenty-five pounds into our boat. The breeze freshened and as we parted, they ran up the German flag at the peak of the gaff and saluted us. Captain Howland did the same.

Tuesday, November 22, 1881

Fine weather and running with topsails double reefed. Mr. Young at masthead sang out, "There she blows, and there she blows, and there goes flukes!" The deck was all alive with men, even the sick man, for that shout puts a thrill into all whalemen from the captain to the cabin boy. A school of sperm whales was sighted on the weather beam. It was a long pull to windward. We chased the whales all afternoon until dusk but could not get near them. At last the flag went up at the mizzen peak calling us on board. It was a long hard pull on the oars in a rough sea, but I did not hear a single complaint from any of the men.

Thursday, November 24, 1881

Thanksgiving day came in with rugged weather and we had to furl some of the topsails. What a great day this is back in New England with its good times and tables full of mince and pumpkin pies! We have not even the goat that we planned on having, but we hope it did the Dutchmen good.

Monday, December 5, 1881

We have had quite a spell of rough squally weather and have not seen a whale. Once we saw a school of killer whales fighting among themselves. Today the weather was better and the ship looks like a washerwoman's back yard. The boys have their clothing hung all over and in the rigging. We all got plenty wet these last few days. This afternoon all hands were looking at a total eclipse of the sun. Some of the Portuguese wondered if the sun was always going to be covered up like that. It got cold during the eclipse and Mr. Reed had to put a sweater on to keep warm. It was a grand sight in a clear sky, not a cloud and nothing to break the view.

Sunday, December 18, 1881

Outside of some finbacks and a few ships, we have not sighted a thing for a couple of weeks. The men have busied themselves making trinkets. At masthead today, I sighted Port Dolphin at the southern end of Madagascar. Every Sunday I have been giving John Brown and Frank Gomez their lessons. Mr. Young wondered if I did not get tired doing it and thought I had a lot of patience, but I told him it was a pleasure. Sam Hazzard asked John what he was going to do when he got back to New Bedford. He said that if he had five hundred dollars coming to him, he would go home to Brava. He has a piece of land there, five acres, that he bought with money from his last voyage. There is a little two-room house on it built of stone, which he has rented until he gets back. He said that there was plenty of wood on the island and that he could live there a long time by selling chickens at ten or fifteen cents apiece, and having a goat for milk, and raising some pigs. He said he was going to buy an old boat, fix it up so he could fish, as there are plenty of fine fish to be caught. He would not need much clothing as there is no cold weather, and there is plenty of fruit of all kinds. John is quite handy and has good sense. He is good and strong and a healthy lad. He is black as they come but as good as any white man.

Monday, December 19, 1881

John Brown got a hold of a story book from Frank Knowles and is head and ears into it. Every few minutes he asks what this or that is. John is a scream when reading as he cannot read without spelling some of the words out loud.

Some of the men are making trinkets out of tortoise shell. Mr. Reed made a pair of bracelets about an inch and a half wide of a real pretty design. He got me to mark the letters M. E. J. on a silver plate. He cut and filed the design and fastened it on with silver rivets around the edge, then polished it. It looks grand.

Friday, December 23, 1881

The cooper made battens for the foretopmast stay and this afternoon I seized them on to keep the jib pennants from chafing the stay. The hands forward are still busy making tortoise-shell watch chains. One Portuguese made one three feet long with an ebony cross at the pennant. You would wonder at the fine work of some of these rough-looking men who cannot read nor write. They polish their work with shark skin until it looks like glass. I made a small kedge anchor of ivory with a stock that would fold. I have also made a small box inlaid with a star of pearl shell, and another one with a star and crescent. Lately there has not been much to do, so all of the men have been working on all sorts of things. I made a pair of small anchors out of ivory for earrings and a breastpin in anchors for my sister Maizie. I took time with them and they look neat and pretty. The shackles for them I made of gold by cutting up a two and a half dollar gold piece. I also made a small ivory anchor for my brother Sandy's watch chain.

Saturday, December 24, 1881

Two of the men forward had a scrap and the mate sent them to masthead for punishment. They did not mind that one bit. One of them took a piece of rope with him, lashed himself to the rigging and went to sleep up there. The mate found out about it and left him up there until ten o'clock at night but called the other fellow down at dusk. They were both satisfied. Take it all in all, the men are agreeable all around and not a bad lot; no fierce fights and do not use knives like some.

Sunday, December 25, 1881

Christmas day in the Indian Ocean. Fine weather, light wind and all sails set. Like some others on board, my thoughts wandered to my far distant home, mother, and her many kindnesses to me. As I look back to the days of my youth, I was a wild, rough, bareheaded boy, but as I think of it I must have been a favorite of my mother. She showed me many kindnesses and would have me kneel down and pray. I cannot forget the entreaties to Him above for protection for her wild and wilful black sheep as she called me. The thought often comes to me of my wild, half-civilized, barefooted tramps through the snow to the schoolhouse back in the Highlands of Scotland.

Today we had the usual Sunday duff, baked beans, bread, salt pork, rice pudding and coffee, but plenty of it. I feel thankful to have good health and a good appetite.

Sunday, January 1, 1882

We are still to the south of Madagascar looking, searching, and wishing for whales. We have all been thinking today of the good times in New England and turkey and pumpkin pie. I told Tommy, Sam, and Mr. Reed of the New Year shooters in Philadelphia, and the parade of clowns and masqueraders with all the comical and fancy dresses, and the floats. They had never heard of anything like it. We talked about the Arctic. I said I liked it but Sam said one dose was enough for him and that he had one tough old time.

Tuesday, January 17, 1882

Today the third mate's pipe fell overboard. We hauled aback, lowered a boat and got it again. It was a fine meerschaum pipe so it floated. He valued it very much and had had it a long time. He got it from a German that he helped around the Galápagos Islands some time ago. We have not sighted any whales except finbacks for a long time. Sometimes these finbacks are very large and go through the water with great speed, but none of the whalers bother with them.

Thursday, January 19, 1882

At eleven o'clock today we had a bad squall. Called all hands to shorten sail in a hurry. They came on the jump and manned the clew lines, buntlines, downhauls, and let go the halyards. In twelve minutes we had everything clewed up but the topsails, and they were down on the lifts and the reef tackles hauled taut. In less than an hour we had all sails on her again, heading and bowling along to the westward, heading towards the Cape of Good Hope and rough weather. This is a squally place and you have to be on the lookout all the time so as to get the sail off the ship right quick, for the squalls come up without warning. You are likely to lose masts and gear if you are caught napping. We pass a lot of vessels now as we approach the Cape.

Wednesday, February 8, 1882

We pumped out the scuttle butt today as the water in it was bad, not fit to drink, and filled it with fresh water. At noon we sighted a small whaling schooner and ran down to her. It was the *Gage H. Phillips* of Provincetown, Mass. We had a gam with her and got some late papers. They reported that they had seen whales yesterday but as the whales were going fast, they had no chance.

Thursday, February 9, 1882

The day came in with fine weather, light trades and set all sail in Lat. 14° 20′ S., Long. 5° 0′ E. At nine o'clock Antonio sang out, "There she blows, there goes flukes." All four boats chased the whales going to windward. We pulled hard until three o'clock but could not catch up, and finally lost sight of them. We got back to the ship at five o'clock, tired and hungry. No sooner were we through with supper when someone spied a lone whale coming along to windward. Again we lowered four boats. Mr. Young got fast at six o'clock and had a tough time killing it in the dark. You could hear it before you could see it, snorting like a steam engine. Mr. Reed finally got a chance to put a bomb into it. That fixed it but we had to stern all in a hurry when it went into a flurry. We got the whale to the ship and fluke chains on about eleven o'clock. Chasing whales at night is not the safest job I know. This whale gave us seventy barrels of oil.

MOZAMBIQUE TO ST. HELENA

Thursday, February 23, 1882

My old friends the tropic birds were the only things in sight while I was aloft today. They keep piping and flying around your head. Sam Hazzard is making a jigger wheel for pastry out of a whale's tooth. Tommy Wilson is making a bone cane with an ivory handle and Mr. Reed a rosewood cane with a handle of a man's fist grasping a snake with its head stuck out and its body wrapped around the wrist. It's a fine job and looks wonderful. Some of the men are sawing up the jaw pan of a whale's jaw for bone canes. Three or four are making coconut dippers with different kinds of handles. Some break the shell if not careful when riveting the handle to the bowl. For the bowl, wet shark skin is used for polishing but for shell, ivory or ebony, a revolving wooden wheel is used, giving a polish as smooth as glass.

Tuesday, February 28, 1882

There were two ships in sight at daylight today so we ran down and gammed them. One was the whaling barque *Pioneer* of New Bedford. She is one year out and has 240 barrels of oil. This old barque has straight sides, rather narrow, no sheer, and stands high out of water. She's a real old timer and while she cannot sail very fast, is a dry sea boat in a gale of wind. The other ship was the whaling barque *Wanderer* of New Bedford, a fine barque, almost new. She is larger than the *Kathleen* and is fitted with everything of the best.

Sunday, March 12, 1882

For two weeks we have been cruising for whales with no luck, and are now heading toward Jamestown to ship oil home. As we approached St. Helena, we saw quite a few ships. Yesterday, we gammed the barque *Petrel* of New Bedford. Today, as we cruised to windward of the island, the whaling barque *Greyhound* of New Bedford ran down to us for a gam. We saw plenty of sail going past the island signaling to the signal station. There is no place to land but around at the lee side at Jamestown valley. It's a beautiful, healthy, pleasant place to live.

Geraniums grow wild all over the hills, also prickly pear, a cactus, the fruit of which is good to eat.

ST. HELENA

Monday, March 13, 1882

After anchoring in forty fathoms of water off Jamestown, we furled sail, got the running rigging off the deck and stowed up on the tops and sheer poles out of the way. We need every bit of deck room to get the casks of oil ready for shipment back to New Bedford. Each cask has to be examined and coopered. We got a spare boat down from overhead for a shore boat, to be used for taking the men ashore so as not to abuse the whale boats. When the boat got back with letters, there was one for me, the first I've had for a year and it was welcome. The letter was from my mother to whom my thoughts often turn while far away out here.

Wednesday, March 15, 1882

The cooper and a gang of men have been very busy heading over the casks that we have been getting out of the hold up on deck. We hove short on the anchor and got close to the schooner *Lottie Beard* which is going to take our oil back to New Bedford. Hoisting the casks out of the hold is hard, dirty work and will keep all hands busy for a few days. There are several whaling barques at anchor here, the *Pioneer, Sea Ranger, Petrel, Wanderer, Morning Star,* and *Greyhound*.

Thursday, March 16, 1882

I went on shore tonight and met two of the Ninety-third Highlanders and some of the Ninety-first Gordon or Seaforth Highlanders. I asked them the time of day in Gaelic. At first they thought I was a Portuguese, but when they heard me speak Gaelic, they could not do enough for me and asked me to come up to the barracks on Ladder Hill to meet some more Highlanders. I promised I would come. I stopped in at the Port Officer's place to bid him well and then went up to see his family. Mrs. Jamieson said, "You have been here three days and why did you not call in?" I told her that this was the first I had been on shore. She asked me to come to dinner on Sunday and spend the day. Then Mr. Jamieson came in and nearly shook the hand off of me. He wanted to hear about the sights I had seen around the islands in the Indian Ocean. We had some songs and then I had to get back on board.

Friday, March 17, 1882

A couple of Highlanders came out from shore looking at the stern of each vessel trying to find the *Kathleen*. I said, "Who are you looking for?" in Gaelic, so they came aboard. It seems they were anxious to see a whaler. I showed them all around, not only the gear and the boats, but explained how we cut up a whale. We had a good time talking about the Highlands. Mr. Young asked me and my friends to come down in the cabin and have supper with him. He gave each of them a whale's tooth for a souvenir. When I pulled the men on shore, they asked me to come up to the barracks on Saturday night for a dance. I was to ask for Mr. MacGregor or Mr. MacDonald of Skye.

Saturday, March 18, 1882

While the men were painting the outside of the ship, Mr. Gifford wanted me to paint the name on the stern and said I could go on shore when I got through. I told him that I would like to stay until Monday morning and he said, "All right, Bob, go ahead."

When I got on shore, I went straight up the street to Lady Ross' house. She gave me a great welcome and threw her arms about me and kissed me. She is certainly a wonderful old lady, living here all by herself, to watch over the grave of her husband, away from all her people and with only two servant girls in the house. I told her that I had been invited to the barracks for a dance. She said, "I suppose you have had no dance for some time, nor any girl. You would not dare to ask any of the Jamieson girls, for their mother would not let them go up there to a dance, but you will find lots of girls, some of them wild, so be careful. I will wait up for you."

It was after eight o'clock when I got to the barracks. All were soldiers, except four civilians, but there were lots of girls. I asked for Mr. MacDonald and Mr. MacGregor. After they met me, they introduced me to their lieutenant and their captain, both of whom talked Gaelic to me. The captain got me a partner, a good dancer, but she smelled strong of whiskey. I danced one set with her and that was enough. Another set was forming and a young woman said, "Come on, Yankee sailor." She too could dance but smelled of rum. Mr. MacDonald then got me a fairly good partner and I held on to her all evening. When leaving, I shook hands with Captain Ross, told him that this was the first dance that I had had in three years and that I had enjoyed it. He said he would let me know when they had the next one. I found Lady Ross waiting up for me and we talked until one o'clock about the Highlands and people we knew.

Sunday, March 19, 1882

I told Lady Ross about the first partner I had at the dance. She said that this girl was a character, who came from an honorable family and had had fine learning but that she was a bad woman and always drunk. Lady Ross and I went to church arm in arm. After church, we were surrounded by the Jamieson family, and had lunch with them. I told them about our voyage in the Indian Ocean. Lady Ross asked Jennie, the eldest girl, to come along with us up to her house. While the old lady had a nap, Jennie and I had a good talk. After we had tea, I took her home and came back to Lady Ross' for the night.

Tuesday, March 21, 1882

Yesterday we were busy all day breaking oil out of the hold. There was another gang over on board the schooner stowing the casks away. Today we were still at it but we knocked off work about four. I went on shore and up to the barracks, climbing the steep stairs, eight hundred steps or more, cut in the face of the rocks. When I got to the top, there was a large, smooth, rock plateau about four hundred feet long and two hundred feet wide on which were about fifteen guns chained to ringbolts. Looking down at the ships in the harbor, they looked like lifeboats and the men like flies. At the back of the plateau, the rocks rise about 150 feet perpendicular. Against this backing, the barracks run the full length of the plateau. An officer came over and spoke to me. He showed me all over the place. I had supper in the officer's mess hall after which there was some fine singing and playing on the flute and violin. As the gun boomed the warning to get on board, they all pressed me to come again and tell them more about whaling.

Friday, March 24, 1882
Finished taking oil out of the hold today. The hold needs some straightening out, for everything is topsy-turvy. The ground tier will have to be re-tiered and filled with water. I went on shore tonight and met Miss Jennie Jamieson, took her for a walk on the promenade and in the Governor's garden. She wanted me to go up to the house with her. I told her that I had had a hard day of it and how the casks of oil had to be taken up and sent home on another ship because ours could not carry it all. I told her I might be on shore on Sunday again if I could get off. She said, "I'll meet you at the church door and we'll take a walk." I told her about the dance up at the barracks and about the women I had danced with. She said that her mother would not let her go up Ladder Hill. I told her about the women smelling of whiskey and she said that no nice girl would go up there and why did I go. I said some of the Highlanders had asked me to come. The warning gun boomed and we had to say good night. What a contrast is this fine girl to the drunken girls up on Ladder Hill.

Saturday, March 25, 1882
All of the oil is out and on board the schooner *Lottie Beard*, 982 barrels. A good catch for the year, around the Indian Ocean. Cleaned ship, scrubbed down good fore and aft, and all done at noontime. Sam Hazzard and I shaved and got on some good clothes. We had permission to go on shore until the next

night. Three of us harpooners, as soon as we were on shore, went straight through the town and away up the valley. It was beautiful with geraniums in full bloom. There was a stream of clear cool water which fell in a waterfall over the edge of the rocks two hundred feet below to supply the town with water. We went up a stony but shaded road, one side of which was covered with cactus. At the top, the road ran on to a plain, past a few homes, then on to Longwood, the house where the great Napoleon lived and died when he was a prisoner here. Coming back, we came to Napoleon's tomb, near which is a spring of water with weeping willows shading the spot. When we got back to town, I was wondering what to do, when along came Lady Ross and that settled it. I went home with her and told her about our walk to Longwood and Napoleon's tomb, and about promising to meet Jennie Jamieson at church tomorrow morning. She wanted to know if I had a girl in America. I told her that I had never been ashore long enough to become acquainted with any. After we had supper, we talked until midnight about places and people we both knew. She sang Scotch songs which brought the tears to my eyes.

Sunday, March 26, 1882

I did ample justice to a fine breakfast of porridge, bacon and eggs, fresh rolls and coffee. A little after ten o'clock, Lady Ross and I went down the road to the church. On the way we passed some of our crew and Captain Howland. Lady Ross said that she knew our captain. After the services, Miss Jamieson and her father and mother came over to me and made me promise to come down in the afternoon and have supper. After dinner, before I left to go to the Jamiesons', Lady Ross told me that any time I was on shore I was to make her house my home, that she would be a mother to me and that I gave her great comfort. I think a lot of this dear old lady because of her kindliness to me.

The Jamiesons wanted to hear all about our cruise in the Indian Ocean, so I told them of Madagascar, of Bird Island, the dhow slavers and the British gunboat *Sea Gull*. After supper we had some hymns. Then, when I was in the midst of telling about

MOZAMBIQUE TO ST. HELENA

whaling, the warning gun boomed and I had to leave these good friends and hurry on board the ship.

Tuesday, March 28, 1882

Yesterday and today some of the men have been giving the ship another coat of paint, but most of them have been helping the cooper and fitting in casks in the main hold. We got all dirty and oily but had a good scrub afterwards in the deck tub. Sam and I scrubbed one another's backs. After supper I wrote some letters home to Philadelphia. Some of the men went gamming on the barque *Merlin*, which just came in. There are a lot of ships in the harbor: *Milton, Eliza Adams, Niger, Desdemona, Morning Star, Sea Ranger, Falcon, Petrel, Sunbeam, Hercules, Stafford, Ocean Wave, Mattapoisett*, and *Mars*. All the time I was writing, Sam Hazzard was telling about his girl in Attleboro and then he began to bawl out and sing, "Take This Letter to my Mother." I was tired out and went to sleep with Sam Hazzard's song ringing in my ears.

Thursday, March 30, 1882

Tonight I went on shore and up on Ladder Hill to have supper with the soldiers. After supper we had music and dancing. This time I danced with sober girls, quite a different class from the first time. We had a jolly good time until late at night, the cornet, flute, and fiddle going full blast. During intermission some of the Highlanders sang Gaelic songs, "Annie Laurie" and "Mary of Argyle." I tried to come away two or three times, but no; I had to sleep in the barracks all night and have breakfast with the Ninety-third Highlanders.

Friday, March 31, 1882

After coming down the steps this morning, there was our second mate, Mr. Young, and a mob of soldiers and sailors on the corner. Mr. Young told me that a big burly soldier, one of the Seventh Engineers, and he were going to have a fight all over a trollop of a woman. The soldier wanted to beat her up and the second mate would not let him touch her. Mr. Young said, "Bob, stand by me." I said I would. Mr. Young sailed in and they went at it rough and tumble, kicking and gouging. All brute strength, no science; everything fair, nothing foul. The soldier got Mr. Young down but with a quick movement Mr. Young turned him over and sat on him. He pounded the soldier on the face and bunged up both his eyes. The claret was flowing, for the soldier's face was all raw from Mr. Young's fists. A policeman who had been looking on all the time and had been laughing and grinning when the soldier seemed to be having the best of it, stepped up and was going to swing his club on Mr. Young's head but I grabbed the club and took it away from him. As I did that, I sent him spinning head over heels. Two of the Highlanders took the policeman and ran him out of the crowd and told him they would lock him up if he bothered any more. Mr. Young pounded on the soldier's face until he begged for mercy. When all was over, the soldiers told me they were glad Mr. Young had given the drunken bully a beating. I spoke to the policeman whose club I had taken, saying that I did it because with all the ship's crew gathered there, a riot might have been started if he had used his club on Mr. Young. The policeman was not a bit angry. If a riot had been started, there would have been murder and bloodshed, for let me tell you these sailors know no fear, and as they all carry sheath knives and know how to use them, somebody might have been killed. I got on board late and expected to catch the devil, but I heard no more of the fight.

Saturday, April 1, 1882

Mr. Young did not come on board last night and Mr. Gifford, the first mate, was angry at everybody as he had Mr. Young's work to do as well as his own. Mr. Reed did not show up either. Today the captain came on board and asked where Mr. Reed was. I had to tell him that he had not been on board for two days. The captain also asked me about the fight Mr. Young had and I told him how it all happened and came about. He told me that he had given Mr. Young his discharge yesterday and that he was going home on the *Lottie Beard*. The captain told me to look after things, for I was the only one aft, until Mr.

Gifford got back on board. There are a lot of recent arrivals to the whaling fleet in the harbor: *Lane, Tropic Bird, Rose, Bertha, Gage Phillips, John Carver,* and *Andrew Hicks.*

I cooked a good supper for the men: chicken pot pie, roast sweet potatoes, hot biscuits, pudding, and custard pie. It tickled the palates of this happy bunch of boys, eating on deck and laughing all the time. Mr. Gifford signaled for a boat at eight o'clock. I called the boys and they came on the jump. When the mate came on board, he said that he was hungry, so I fixed up a nice supper for him. Mr. Gifford said that I surely knew how to cook. He also said that the captain told him how I stuck by Mr. Young in that fight on shore and how I could be trusted at all times. I told Mr. Gifford that I had good friends among the soldiers and had no reason to be afraid of them.

Wednesday, April 5, 1882

As I was not likely to get on shore again for some time, I went this morning. I did not know whether to go up to the barracks, to Lady Ross, or to the Jamiesons. I have to bid them all good-bye. Although a stranger to them, they have all made me feel at home. Just then I saw three of the soldiers doing guard duty at the gate and said good-bye to them, for I might not see them when we get back here again. Along came Mr. Jamieson, who said, "Are you sailing soon, Bob?" I replied, "Most likely tomorrow." He linked his arm in mine and said, "Come on up to the house, for you are not likely to be on shore again." I had a bite of dinner and a talk with his wife and daughters, for Mr. Jamieson had to go back to work at the jetty, being Port Officer. I finally said that I ought to go up the street and bid Lady Ross good-bye, but they wanted me to wait until after tea, when one of the girls would go up there with me. Mrs. Jamieson said, "I know when you get up there, she will not let her Highland boy get away until morning." I unwrapped the two coconut dippers and handed one to Mrs. Jamieson. They all admired the polish, the rosewood handles, and wondered how I managed the wreath of the walls of Troy. Mrs. Jamieson knew the other dipper was for Lady Ross because of the thistles all around it and remarked of the bonnie black handle and on my having done such fine work. After tea the oldest girl, Jennie, came along with me and we took a roundabout way. We had a good talk about America and my folks. I told her that I might not be back to St. Helena for some time and she said she wished I could stop on the island, but I told her I could not leave the ship until I got back to New Bedford. It was an hour before we got to Lady Ross' although I have walked it many a time in fifteen minutes. When Lady Ross saw me, she kissed me like a mother would, God bless this dear old soul. Jennie Jamieson only stayed a short time, and when I went to the door with her she looked around and, seeing no one, reached up and kissed me, then quickly ran away, saying, "Good-bye, Robert, I will watch for you in the morning when you are going on board." Well, that was so unexpected I never thought of anything, no time to think or know what to do. It was done and she flew away like a flash. I came back into the house and listened to Lady Ross talk of the time when she was a girl. I asked her to sing and she sang two songs that I had never heard before about Bruce at Bannockburn.

Thursday, April 6, 1882

After breakfast, when I said good-bye to Lady Ross, she gave me her blessing. I got on board at eight o'clock and all of us were soon busy getting ready for sea again, getting the rigging down on deck and in place on the belaying pins. Mr. Young came on board for his sea chest. When he saw me, he grabbed both my hands and said, "Bob, you are a man, and if you ever come out to the Sandwich Islands hunt me up and stay with me. You have been good to me and I shall miss you and often think of you. Good-bye."

Friday, April 7, 1882

Mr. Gifford said, "You four boatsteerers have worked hard and all four of you can go on shore this last night but you must come on board at nine o'clock." The men forward refused to put us on shore and the mate went to the forecastle, called all hands and made them scrub deck for two hours. I went up the street at dusk and met Jennie Jamieson. She said that she came looking for me because her father had told her that the *Kathleen* was sailing tomorrow. We went into the Governor's garden and sat on a bench looking at the ships at anchor in the harbor. She kept close to me all evening, and when the warning gun boomed, she clung to me and would hardly let me go. It was hard to part and say good-bye, but when I saw the men on the jetty I had to go quickly. When I got on board, I thanked Mr. Gifford for letting me go on shore to say good-bye to my friends.

FROM ST. HELENA TO KABINDA, CONGO, AFRICA

Headed towards the coast of Africa—Humpback whales—Nantucket sleigh ride—Sunken whale—School of sharks—Visited by blacks—Congo River—Native chief—On shore at Kabinda—Hunting trip—Lost in the jungle—A large serpent—Aided by natives—Deer—Sign language—Treed by a lion—Native graveyard—Back to the ship—Lectured.

CHAPTER VII

FROM ST. HELENA TO KABINDA, CONGO, AFRICA

Saturday, April 8, 1882

This day came in with fair, clear weather and light winds. After breakfast, all hands were called to get ready to sail. We hove short on the anchor and sheeted home the fore and main topsails. The topsails were hoisted and we manned the windlass with the chantey "Good-bye, Fare You Well." We are laying off and on as the captain is still on shore. We have a new second mate in place of Mr. Young, who got his discharge. The new man is a Portuguese, a Mr. Roderick. Mr. Reed, the third mate, quit the ship. Besides him, there were four men who left, but we got four new ones in their places. We hoisted the shore boat up on deck and stowed the oars away until we get to the next port.

Sunday, April 9, 1882

It is a glorious day standing up here on the royal mast searching for the spout of a whale. We are a little to windward of St. Helena and there are lots of ships going past the island with strings of signal flags fluttering in the light trades as the signal station answers. Mr. Gifford said, "Bob, that was a good-looking girl who came up to you. Who was she?" I told him that she was the Port Officer's daughter, a fine girl and as good as she looks.

Monday, April 10, 1882

We are still laying off and on the island. At noon Mr. Reed came off for his things and his chest. He got his discharge too. He went off without saying good-bye to me. He is so infatuated with some girl that he forgot that I was his harpooner and struck all of the whales for him. The captain came on board at five o'clock and gave tobacco to all those who wanted it. All hands feel as though they had a good time at St. Helena, as Tommy said, "A sailor's heaven." Tommy said that he had heard that I was at a dance at the barracks. I said that I was and that the women smelled like rum shops but were good at dancing. He said that one of the women I had danced with told him about it, and that I was quick on my feet but too proud.

Tuesday, April 11, 1882

The fourth mate, Mr. King, has now been made third mate and will take charge of the bow boat. Sam Hazzard has been made fourth mate and will head the starboard boat. Sam took James Lumbrie, the little red-headed boy, for harpooner to steer him. The captain told me to go in the waist boat with Mr. Roderick as his harpooner. We did not shorten sail tonight and are working toward the west coast of Africa, where we are going after humpback whales. This is a different whale than the others I have helped catch and I understand that the method of whaling is quite different.

Sunday, April 16, 1882

Today the men are reading papers that we got at St. Helena. We had no time to read them while there. The soldiers gave me some Glasgow papers and a bundle of *Illustrated London News*. Some were six or eight months old, but they were new to me. Mr. Roderick is looking at the pictures and asking all kinds of questions about them. He cannot read and I have to explain everything to him. He seems to be a right good sort of a fellow and he said, "Bob, we will have to pull together."

Sunday, May 7, 1882

There has not been much to do for three weeks, outside of regular duties and repairing the rigging and patching the sails. We have not sighted a single whale except finbacks, and we do not want them. I held school for a little while today. John Brown has done very well in a little more than two years' time, only getting a lesson once in a while. I feel glad that my two boys have done so well and that I took the time to teach them. Frank Gomez got a letter from his girl while in St. Helena. He managed to read it all by himself and today was the first chance he had to let me see it. She wrote a right nice and very plain letter and told Frank to thank me for my kindness in teaching him to read and write.

Saturday, May 13, 1882

The cooper has been making new boat kegs. These hold about three gallons of water and are supposed to be kept filled with fresh water all the time. Three gallons for six men is not much when pulling an oar all day in the broiling sun. The cooper also made some small bailers with side handles, to bail the boat and for wetting the whale line to prevent it catching on fire if it starts to smoke as it runs around the loggerhead. Had a gam with the full-rigged whale ship *Eliza Adams*. Her captain stayed on board of us all day. The harpooner that came with him was a big powerful negro, as black as could be, and stood six feet four inches in his bare feet. He came from Dominica, one of the West Indies, and spoke good English.

Mr. Roderick asked how I taught Frank to read as I was not a school teacher. The second mate cannot read nor write. He speaks good English, knows his name when he sees it, can tell time by the clock, but does not know the letters of the alphabet. I said that he ought to learn to read, but he said that he did not care to, so I did not urge him.

Monday, May 29, 1882

There were lots of fish alongside the ship today. The men caught a tubful, some large, some small, of many different kinds. From all appearances, we must be getting near the coast of Africa, for there are lots of birds flying all around. Some are boobies that dive from great heights so swiftly that you would think they would break their necks, but they always get their fish.

Friday, June 2, 1882

Yesterday we got the anchor to the cathead all ready to let go. At nine o'clock we sighted land but it was kind of hazy and hard to see this low-lying land. At eleven o'clock we clewed up sail and let go the anchor in sixteen fathoms of water. The rest of the day was spent in getting everything ready for humpbacking, even to the pots and mincing machines. Today we called all hands at daylight, had breakfast and lowered all four boats. With a good sea breeze, we sailed down the coast looking for humpbacks in from five to twenty fathoms of water but did not see any. We had lunches in the boats and got back to the ship at six. These whales come up the coast from the south to calve in the warm shoal water along the land. The water along this coast is alive with fish. You need not bait the hook, just throw in the line and haul in a fish every time.

Sunday, June 4, 1882

At daylight we hove up the anchor and started to the north. Sam Hazzard raised a humpback whale after breakfast. We lowered all four boats and chased it until after two o'clock. I got a chance and darted my first iron into a humpback. By the time we had the mast and sail out of the way, we were traveling some, with the water flying from our bow, and making for the south west as hard as we could go. Mr. Roderick shouted for me to come aft and take the steering oar and to tell the men to haul line as I got aft to the loggerhead. The whale milled around and came straight for the boat, but I slewed her around with the big steering oar. After hauling line, I threw the bow of the boat on the whale close to its fin so that Mr. Roderick got the lance into a vital spot. The whale churned, spouted clots of blood, and rolled over, dead. It took us until five o'clock to get alongside of the ship. The humpback is different from other whales. It does not have teeth like the sperm whale but has a bone sifter like the right and bowhead whales, only the bone is shorter. Its belly is like an accordion with grooves running lengthwise. There is a very long fin on each side, six to ten feet long by about a foot broad, but only two or three inches thick. It can cut a boat in two with these fins. When you dart a harpoon into its body, the first thing it does is go down to the bottom, then it comes up and goes away as fast as it can go, dragging the boat after it with the speed of an express train. That is what whalemen call a "Nantucket sleigh ride." It has to

TO KABINDA, CONGO, AFRICA

be quick work to get the mast down and sail rolled up, and at the same time keep clear of the whale line that is humming out and smoking around the loggerhead. When everything is all clear and out of the way, it is "Haul line, my hearties." All hands pull hard on the line to try to get the boat close enough to the whale to lance it. The officer and the harpooner have to keep their eyes open and be ready to back away quickly. When the whale runs offshore into deep water, the lines from both tubs may run out. This being the first whale of the season, I got a bounty of five dollars.

Tuesday, June 6, 1882

Yesterday, after lowering the boats, we pulled to the south where we sighted whales. We chased them all day but never could get close enough to them to strike. Today one whale ran first into shoal water then turned and ran offshore, just playing with us. All day we pulled in the hot sun after it, but could not get near and came back to the ship at sunset, tired and hungry.

Wednesday, June 7, 1882

One boat's crew stayed on board today to help boil oil, but the rest of us pulled to the southward. We saw no whales but sighted a small trading steamer from Liverpool that was anchored close inshore at a trading station. We went on board and had a talk with the men. I talked to the quartermaster in his quarters, which is a regular store, having clothing, hats, suspenders, brass buttons, knives, and all that sort of thing to trade with the natives.

Saturday, June 17, 1882

Every day we have put out for whales but have had no luck. Some days we could not see any and other days we would chase them but never could get close, and we would come back, hot, tired, and hungry. Today, just before sunset, Mr. Gifford got fast and put a bomb into one and stopped its monkeyshines. That killed it but it sank. All four boats tried to pull it up but could not budge it, so we left the starboard boat anchored to the whale all night.

Sunday, June 18, 1882

We went back to where the whale sank and tried to pull it up and could not. The sharks had eaten the whale all up but the bones. They are as thick around here as minnows and travel in schools up and down the coast. A boat from the barque *Andrew Tucker* came over this afternoon for a gam.

Friday, June 23, 1882

Everything on board this morning was wet from the heavy dew. Whales were sighted and the third and fourth mates gave chase. We, in the waist boat, were eating lunch. I stood up waiting for one whale to come up. I could follow it all the time for the water was shoal. When it came to the surface, it was close to the boat. I was standing ready with the harpoon in my hands. Mr. Roderick threw the bow of the boat toward the whale, giving me a good chance to dart both irons in pretty well forward, which allowed us to keep up close so that he could get at it with the lance. It soon rolled over dead, but the crew had to be on the alert when it went into a flurry. The other crews went back to the ship, got her under way and alongside of the whale at eight o'clock.

Saturday, June 24, 1882

Not a breath of air today. The water was covered with birds and swarming with fish. Two boats went after whales, but our crew stayed on board to help cut in yesterday's whale. While cutting in, a lot of natives came alongside in dugouts, each holding one man. The captain cut chunks of whale meat for them, all that their canoes would hold, and also killed half a dozen sharks for them. The captain gave them some old pieces of rope to tow the sharks. We lay about three miles offshore in eight fathoms of water. Some of the natives had so much meat in their dugouts they had to get in the water and push by swimming back of their boats. These natives are large, finely built men,

TO KABINDA, CONGO, AFRICA

stark naked. They have no fear of the sharks and kill them by going right up and hitting them across the nose with a billet of wood. There must have been at least thirty-five natives who kept laughing and singing on their way back to the shore. I could hear them until they reached the beach and then there was one grand loud shout.

Sunday, June 25, 1882

This is a day of rest. The weather is fine but no wind and hot. Some of the men are fishing in water so clear that you can see the bottom eight fathoms below. There must be fifty or sixty natives surrounding the ship in their dugouts; black men, black as the pot, laughing and talking all the time. The carcasses of the whales sunk alongside bring sharks here in schools. Sam Hazzard has killed several for the natives and once in a while they get one. The sharks are all towed ashore, where they are hauled up on the beach and covered with seaweed and rocks. The natives build a fire on top and roast them. These natives are apparently a happy lot, powerful, well-built black men, with not even a handkerchief to cover them. Several of the blacks came on board to trade grass mats, sharks' teeth, and other things. Our boys had a lot of fun with them. Sam Hazzard traded a red undershirt with one of the natives who put it on and strutted up and down the deck. It was the only thing he had on. Late in the afternoon the natives made for shore with so many sharks that they could not make much headway.

Monday, July 3, 1882

Last week we chased whales every day in the hot sun and only got two which were not very large and did not give us any trouble. I got a lively one today that we had chased for a long time. I took a long dart at it but fetched it. The old mate sang out, "Good boy." As soon as I got the mast and sail clear, the battle began. It was in shoal water. How that humpback lashed its tail and dove time and again! It began running around in circles and then kept rolling over and over. I got back to the steering oar and soon had the nose of the boat up to the whale, where Mr. Roderick kept churning it with a lance. The boys had to back away in a hurry and give slack line. Finally it rolled over on its side and all four boats towed it to the ship.

Wednesday, July 5, 1882

Got another whale today. This one was wild and ran all the way up and down the coast even though two boats were fastened to it. It seemed like nothing for it to pull the two heavy boats. First it would run offshore and then inshore. Once it came near running into the ship and then we had to slack the line some. These humpbacks are not very big and give only about twenty-five barrels of oil to the whale but we see plenty of them nearly every day. It's hard work and they give us a merry chase every time. After cutting up these whales the decks become very slippery and slimy with "gurry," not like cutting in the clean sperm whales.

Saturday, July 8, 1882

We got one whale Thursday and another one yesterday. This day came in with cloudy weather at daylight. All four boats sailed to the south. When it cleared up whales were sighted on every point. The starboard and bow boats each got a whale. After we got them alongside, there was such a large school of sharks tearing at the whales that we had to lower the landing stage so that by standing there with a spade cutter we could drive them off. I killed about thirty, cutting some almost in half, with their entrails trailing. The other sharks would immediately start eating those that were cut. These wolves of the sea were just as bold as ever, even with their tails cut off.

These have been busy days lately, what with chasing whales, cutting in, boiling down and stowing the oil away. The deck was so cumbered up with casks that we had to get it clear. If we run the oil down to the casks in the hold with a hose, it does not take long.

Wednesday, July 12, 1882

We must have sailed fifteen miles in the boats today before

we saw any whales. We chased three but only the larboard boat got fast. All four boats surrounded it trying to lance it. As it went into a flurry, the boats had to scamper out of the way for safety. The fourth mate's boat was too close, for when the whale rolled over, its fin struck an oar and broke it to splinters.

Saturday, July 15, 1882

This morning we chased whales for a while with only three boats on account of one crew being needed on ship for the whales we got Thursday. The ship hove anchor and ran ten miles to leeward. We sighted a whale all of a sudden right ahead of us. We followed it and although it swam under water, we could keep it in sight all the time, as the water was shoal. When it came up our boat was right alongside of it. I drove an iron into it. As it dived, it raised its tail and half filled the boat with water, drenching the whole crew to the skin. Up it came running toward the ship with us hanging on to it two boats' lengths astern, trying to get close enough to lance it. Down it went again in shoal water and ran along the bottom. It came up and slacked line and pace not far from the ship. I saw a chance to change places with Mr. Roderick and get the steering oar. When I set the bow of the boat close to the whale, Mr. Roderick lanced it quickly three or four times. I saw it was going into the death struggle and bawled out, "Stern all and slack line!" We got back just in time to get out of reach of its ponderous tail and to watch the circus. The whale certainly thrashed the water in all directions before it rolled over, dead. We find with humpback blubber it is better not to boil it when green but to let it stand a couple of days to get ripe. This oil is not at all like the oil from sperm whale blubber.

Saturday, July 22, 1882

For the past week we have seen whales every day but one. We chased them all day long but could not get close. Today all four boats were called on board and we hove anchor. It was not long until we crossed the mouth of the Congo River. The current at the mouth of the Congo runs swift away out here and the water is muddy. You can see eddies from the current here and there and maybe a tree or a few logs drifting out to sea. The mouth of the river is quite wide and deep.

KABINDA

Sunday, July 23, 1882

Today we headed inshore and dropped anchor in four fathoms of water in the mouth of a small river at a place called Kabinda. On one side of the river is a large trading post and on the other quite a village of bamboo huts. The anchor had hardly been let go when the natives came out all around the ship in their dugout canoes. They were all anxious to trade bananas, oranges, and other fruit. A few of these natives could talk English very well and most of them understood Portuguese. Their canoes are made from a large tree trunk, shaped at both ends and dug out in the middle. Some are four feet wide and twenty to thirty feet long. One came alongside with a great chief in it, all decorated up with a battered high hat on his head. The band on the hat was a red ribbon and on the side was the wing of some brilliant bird. His coat was a British soldier's red coat with a pipe-clayed belt. For pants he wore a pair of woman's drawers with ruffles on them. He had on a pair of large gum boots with brass spurs on his heels. He wore a collar that once was white but now was far from being clean. There were medals of brass and silver on the breast of his coat. On the shoulders, hanging down in front, were old-fashioned epaulets. He was the big chief and would allow no one to board the ship until he let them. Captain Howland invited him on board. Evidently they had met before because the chief said, "How do you do, Captain Howler." They had quite a talk. For port charges, the captain paid him in tobacco, rice, and a few yards of very loud and thin calico, a red, black, and yellow print that only cost three cents a yard back in New Bedford. All he got did not amount to five dollars' worth, but he went away well pleased. We were granted the freedom of the place, permission to trade with the natives, cut wood, get fresh water, and anything we wanted to do without further charge and without being bothered by the natives. I went on shore to look around and found it a nice clean place kept in good order. The huts were made of bamboo and the fronts of them were swept clean and flattened down. The beds consisted of four posts driven into the ground with grass mats stretched between. They use no covers when they sleep, as it is never cold. Their food is cooked in a large cast iron pot that hangs on a tripod. Chickens were running all around, with parrots among them, just like pigeons at home. Some of the natives have pigs, small black ones, and some have a few ducks that are fed mostly on peanuts, which seem to be plentiful. Although the chief was dressed so elaborately, you must not think his subjects dress up very much. The younger ones mostly wear a shadow and the older ones not much more and with no thought of evil.

Monday, July 24, 1882

Kabinda, where we are anchored, is a small place about thirty miles north of the Congo River. There is nothing doing but trading with the natives for fancy grass mats, parrots, monkeys, chickens, and small pigs. Sailors will make pets of almost any living thing, but they all detest snakes of any kind.

Sam Hazzard came to me and said, "Bob, let's go gunning." He asked permission to go ashore. The captain said that the two of us could go as we had worked hard, but to be sure and be back by dark. Sam got a musket down out of the rack in the cabin. The captain said that I could take his gun, a splendid breech-loader, and gave me fifteen cartridges. We took a lunch of some hard-tack and some fresh fried fish. Tommy Wilson put us on shore at nine o'clock. We went through the woods and into the thick forest. There were lots of fine-looking, richly colored birds that had a peculiar shrill cry and whistle. We walked in all directions, and feeling hungry, sat down on the trunk of a tree by a pool of water to eat our lunch. Sam said that we had better be getting toward the ship but we could not tell the north from the south or the east from the west, because the foliage of the forest was so thick. We had no compass and we could not see the sun. As we could not tell which way to go, we got all tangled up in our bearings, so Sam climbed a tree to see if he could see the ship or the water, but there was nothing but trees everywhere he looked. We found some oranges and bananas and sat down and ate our fill. Feeling tired, we scraped some brush together and putting plenty of dry leaves over the brush, lay down with our guns and soon fell asleep, not lying awake to listen for the other's snoring.

Tuesday, July 25, 1882

We awoke quite refreshed but hungry and decided to look around for an African breakfast. We came across two natives and tried to talk to them, first in English then in Portuguese and even tried a little German, but to no avail. We tried to make them understand by signs. They made motions for us to follow them. One took Sam by the hand to lead him. We thought perhaps they were going to Kabinda to trade. Both of the natives were naked except for a grass belt around the waist and they carried long wooden spears about eight feet long, sharpened and hardened at the point by burning. As I walked along, following Sam and the two natives, Sam cried, "Look up." When I looked, there was a large serpent directly above. I jumped. I had such a fright that I vomited. Well, I felt awful sick and weak for an hour afterward. I don't know how long the snake was, but it was as thick as a stove pipe. The natives paid no attention to it and seemed to have no fear of it. Sam said I was white, and the only time he had seen me scared.

A short time before dark, we came to a village of eight huts, built of bamboo. Something was cooking in the large iron pot in the center of the huts and it smelled good. They served supper to us in gourds shaped like bowls. It was chicken stewed with yams, corn, and herbs but without salt. It tasted fine, for we were hungry. A black lad led us to a hut and pointed to the bed, a grass mat stretched above the ground from four posts. We had no covers of any kind but we were so tired we fell asleep without rocking.

TO KABINDA, CONGO, AFRICA

Wednesday, July 26, 1882

I was awakened by Sam wanting to know if I was going to sleep all day. I washed at a little brook, ate a lot of fruit for breakfast and again tried to talk to the blacks. A deer came close to the huts. The natives tried to get close to it with their spears. I motioned them back and taking my rifle, brought it down with the first shot. They skinned the deer, cut it up and stewed part of it and roasted some in front of the fire on spits of wood. Sam and I cut some thin steaks and broiled them over the fire. We had no salt but the meat tasted good. These people are like children, innocent and very generous. They seem to have a dread of our guns and would not touch them for anything.

Thursday, July 27, 1882

Today Sam and I went hunting around the village but not too far. Another native came here today from some other place. I tried to talk to him but no go. He was bright and not afraid of a gun. It came to my head to draw pictures on the ground with a stick and get his interest. First I drew a bird and then a lion, also an elephant. All of the natives gathered around and chattered. After I drew the head of a man and a house, the native who had come last began to make signs and talk to the other blacks. I wondered if they would understand if I would draw a ship, a river and a village like Kabinda. This new lad showed astonishment and got me by the hand. Next he pulled me to the drawing of the ship and pointed to Sam and me. We nodded our heads. He said something but we could not understand his talk. Then the black pointed to the sun and then to the feed. He lay down as if to sleep, then got up, took me by the hand and led me over to the ship. Sam said, "What in thunder is he up to?" I said, "By this fellow's sign language, it means that tomorrow morning he will guide us to the ship." I suppose that our mates on board ship are wondering what has happened to us, thinking, perhaps, that some wild beast got the best of us.

Friday, July 28, 1882

The natives were up at daylight and got breakfast. Two young natives, each with only a hardwood spear, and Sam and I started out. They made good time walking through the forest and brush. The great trees were so thick that we could not see the sky. We pushed through the foliage until nearly dark, when one of the natives climbed a tree and motioned for us to come up. So up we went and in where thickest and although uncomfortable, we fell asleep.

Saturday, July 29, 1882

In the morning we got down out of the trees and ate some fruit that the young black found somewhere. We did not find any good water to drink until about noon, when we found a spring with fine, cool, sweet water. We took a rest, ate a few bananas and started again on our way through the forest. There was no path, just a tangle of undergrowth, and how these young blacks know where they are going is wonderful. Before dark we climbed a tree again. We had to be careful to pick a place where we could brace ourselves in the branches so that we couldn't fall when we were asleep. Sam fell sound asleep while I was talking to him.

Sunday, July 30, 1882

An awful roar just before daylight awoke us. There was a lion roaring not far away. The two black boys showed fear and went higher up in the tree and spread out where the branches and foliage were very thick. Sam wanted to go down and shoot the lion. Inasmuch as we could not see it, I told him to stay up here where we were safe. The lion came closer. Sam made a noise and the lion came prowling right under the tree and stood looking up at us, with his two front paws on the trunk of the tree. Sam with his musket and I with my rifle got ready to shoot. The natives had their spears ready to dart at it. The lion opened his mouth to roar and Sam said, "Let her go," and we fired together. The lion gave one awful roar, the like of which I have never heard, and he rolled over and over, sprang up, kicked, and then, quivering, rolled over and lay still. I suppose the ball entered the brain. We waited quite a while

TO KABINDA, CONGO, AFRICA

and found when we got out of the tree that the lion was quite dead. There was a bullet hole in the roof of his mouth and another in his eye. With our sheath knives, Sam and I started to skin the beast. It must have taken us over half an hour. It was all the four of us could do to roll him over. We let the two native boys have the hide and they seemed very glad and proud to have it by the way they grinned and made motions. We would have liked to have kept it, but we did not know how to cure it and we knew it would spoil very quickly in this heat.

In the afternoon, the blacks came to a halt. One of them climbed a tree and pointed. Sam went up and shouted that he could see the ship. The two black boys would go no further, so Sam and I peeled off our light jumpers and gave them to the boys. When we gave them our sheath knives, the bigger fellow hugged me and patted us both on the back, talking away all the time in his own language. They pointed out the way to the ship by way of a graveyard and we parted from our black friends. We came to a large circular clearing in which lay dead bodies and skeletons with their feet toward the middle of the ring. It smelled something awful. We circled around until we came to a well-beaten path that led us to Kabinda about two miles off. We got a dugout to go on board the ship.

In spite of a damning and a lecture from Mr. Gifford and a blessing from the captain, we were mighty glad to be on board our old home again, the *Kathleen*. If we had been a day later, the ship would have gone away without us. They thought that we had deserted the ship but I believe they were worried for fear something had happened to us.

FROM KABINDA, CONGO, AFRICA TO ST. HELENA

Menagerie on board—Trading with blacks—Native dress—Studying navigation—Ambrizette—Capsized boat—Lengula—Ivory paper weight—St. Helena—Transferring oil—Visiting friends on shore—Dance at the barracks—Cabin boy in irons—Headed for Tristan—Heavy gales—School of sperm whales—Killing whales in rough weather—Christmas day—Whale head lost—Ship model—Big whale—Lost in open boat—No food—No water—Hunger—Thirst—Picked up by barque Fearless—*Landed at St. Helena.*

CHAPTER VIII

FROM KABINDA, CONGO, AFRICA
TO ST. HELENA

Monday, July 31, 1882
We hove anchor and set sail to the north this morning. The ship looks like a menagerie; large and small monkeys, some black and some grey; fifteen or twenty parrots; a lot of parrakeets, chickens, ducks, and at least six small black pigs. The cooper has a dozen small singing birds. One of the Portuguese has the ugliest-looking owl that I ever saw. The fourth mate has a parrot and so have I. All four boats chased a whale today and killed it in short order.

Sunday, August 13, 1882
Every day we have been chasing whales, but they are still wild. It seems as though they are just playing with us. Only two whales have been brought in since the first of August, yet they have been sighted every day.
Today the barque *Stafford* anchored close to us and the captain went gamming. I worked in my boat, straightening out the gear, put new halyards on the sail, turned the ends of the whale line, checked the water in the boat keg, filled the bread bag full of hard-tack, put the bailer and the lantern keg in their places, muffled and greased all the rowlocks, oiled the heads of the lances and repaired their sheaths, made sure that the lances, knives, and harpoons were keen edged, and checked up to see that everything was in its place and nothing missing. I held school for Frank and John today, almost the first chance I have had since we came to this coast.

Wednesday, August 16, 1882
We chased two whales with all four boats. After a while, our boat got near enough for me to dart an iron into one. Away it went for a good Nantucket sleigh ride, snorting and going like all Hades. We passed the other boats with the spray flying. Sam Hazzard yelled, "What's your hurry?" and threw us his

boat warp as we passed close to him. We towed him along for company. When we got up to the whale, Sam helped to lance it. The whale did some queer stunts with its ponderous tail, swinging it around in all directions. For a while it stood on its head and pounded the water with its tail so that it could be heard for miles. It got a crack at Sam's boat and split a couple of planks near the gunwale. After it rolled over dead and was alongside the ship, the lines parted and the whale sank in twelve fathoms of water. Although we sank a large iron into it, we could not raise it. The sharks began to gather for the banquet. They kept coming like a pack of wolves, biting and tearing at the carcass until the water was all red and strewn with bits of blubber and meat. The sharks had a glorious feast. The seabirds flew thick all around, gathering up the crumbs. I killed seven sharks and cut as many in half, and still they kept coming. You could see their large black fins cutting the water in all directions.

Sunday, August 20, 1882

A large canoe came from the shore today with eight big black fellows. Altogether, they had as many clothes on as you could put in your hat. What little they did have was made of grass. They had mats to trade. I bought seven well-woven fine ones and two rhinoceros-hide whips, all for two brass rings, an old razor, and a package of nails. These fellows are great for bright colors. As for clothes, it is the funniest thing to see the way they put them on. One black got a coat from one of the men and seeing his chum put on a pair of pants, tried to slip his long black legs into the sleeves of the coat. A white shirt is a grand thing to them. If one has on a hat of any kind, or only a collar or a pair of fancy garters or gay socks, he thinks he is all dressed up. One of the Portuguese traded a pair of women's long stockings for a parrot and two mats. This black lad put them on right away and strutted up and down the deck. It was a sight for a dime museum to see and behold. These natives were quite sociable and stayed on board all day. Mr. Gifford and I caught two large sharks that we gave to the fellow who sold me the mats, and he towed them on shore about two miles off. You could hear him singing and laughing, thinking of the feast to come. With the glasses, we watched the natives getting ready to make a fire to cook the sharks. At night we could see them around the fire eating and dancing, and hear them shouting.

Monday, August 28, 1882

The past week was a hard one, chasing whales, plenty of them, but no luck. Mr. King got a fine loggerhead turtle. Today we set sail on the ship and headed offshore and started to get things ready for sperm whaling. Casks had to be taken below and the humpback clatter cleared from the decks. The cutting gear, mincing machine, the rail and bitts, and woodwork were cleaned with good hot water and cinders, to get the gummy dirty grease or "gurry" off. The oily soot was cleaned off the smoked-up spars. Everything was scrubbed and then rinsed down good.

Saturday, September 2, 1882

A strong breeze blowing and all sail set beating to windward. Every day since we cleaned the boat up we have been seeing more humpbacks. Today, when we saw several more of them, we decided to try them again. After a great deal of maneuvering, Mr. Gifford got fast. The whale took a great deal of line, for it sounded very deep. When it came up, all the boats were around it and ready for the picnic. We soon had it fin up and towed it to the ship close by. Before we could get the fluke chains on, the lines parted and the whale sank. Sitting around the vise bench, we got to talking about the whales that had sunk after we got them alongside. Mr. Gifford said he never before knew of so many whales sinking like that. He was disgusted and said that we were going after sperm whales for a while.

Monday, September 4, 1882

For the first time in the past six months I got out my quadrant. The old mate makes fun of it every time I bring it on deck. He said, "Bob, is that the quadrant Noah had in the ark?" Everybody had a good laugh but it suits my purpose and today it is Lat. 5° 45′ S. The men are making sennit and stops and rovings for the main topsail. Got some new oars out of the hold to replace those that the whale broke the other day.

AMBRIZETTE, CONGO

Monday, September 11, 1882

Got up some chain and shackled it to the kedge anchor. Having no luck getting sperm whales, we got orders to get the boats ready for humpbacking, so it's to the coast again. The water is covered with birds and the air is full of them, diving for fish. The water is fairly alive with small fish. The ship came to anchor a little to the leeward of Ambrizette, a Portuguese settlement and trading post of about fifty houses, south of the Congo river.

Friday, September 22, 1882

Humpbacks are very plentiful but very wild. Every day we sight them and give chase but have not been able to get near them for over a week. Although blowing fresh from a sea breeze that made a choppy sea, we lowered three boats at daylight. We up sail and ran to the south, sighted whales and chased several. A whale came up under the bow boat and turned it completely over. We ran alongside, picked up the men and towed their boat back to the ship.

Sunday, September 24, 1882

We did not lower the boats today. Some of the boys were fishing and others killing sharks. After dinner I had a lesson for Frank Gomez and John Brown. The captain went gamming on board the barque *Stafford*. I like to stay on my own ship at night but most of the other men like to go gamming. This afternoon I got Tommy Wilson to help me with my geometry as I am a little deficient in it. Tommy is a good scholar. While on this coast I have very little time for study and so get behind in my navigation.

LENGULA

Monday, September 25, 1882

At daylight we lowered the boats and sailed down the coast. We saw a west-coast English trading steamer from Liverpool named the *Bengula*. She was at anchor at a trading post called Lengula. Mr. Gifford went on board and brought back a sack of onions and two bags of potatoes to take back to the ship. These are the first potatoes we have had for nearly two years. Two men from the trading station came off to the ship, both of them Portuguese, but they spoke fair English. They have been on this coast twelve years but hope to go back home to Lisbon, Portugal, after two more years, when they will be retired and pensioned.

Sunday, October 1, 1882

When I came down from masthead today, I studied my navigation, for I have been so busy that I have neglected it. Mr. Gifford helped me with some of my problems and I was glad of it. In the afternoon Frank Gomez and John Brown had lessons for an hour. Mr. Roderick asked me to read an old *Illustrated London News* to him, in which were pictures of some large new steamers and ships. It gave their tonnage, dimensions, and speed on their trial trips. He cannot read and he wanted to know everything that was said about these ships so that he could tell his Portuguese friends when he showed them the paper. Mr. Roderick is fond of pictures and I gave him a good many that the Highland soldiers had given me the last time we were in St. Helena.

Thursday, October 5, 1882

We have been going over all of the whale boats, renailing and tightening them because they had been nearly shaken to pieces on the wild runs with the humpbacks. Some of the boats leaked badly and new pieces of plank and new rivets had to be put in. After giving them a good coat of paint, inside and out, they were put back on the cranes. I took an observation at noon and worked out Lat. 8° 5′ S. I asked Mr. Gifford if I was near right and he said, "Not far wrong with that old quadrant of yours."

Wednesday, October 11, 1882

Just while we were busy repairing the last of the boats today, Mr. King sang out from aloft, "There she blows, there blows a school of sperm whales!" We had to hustle to get all the gear in the boat. There was no confusion. In fifteen minutes we were all ready to lower, the men in their places at the boat falls, and each harpooner at the bow of his boat with his harpoon bent on to the whale line. Mr. Gifford, with his big lugsail, was soon on one whale's back. When it came up, our boat was alongside of it and we fastened to it also to make sure of the whale. Mr. Gifford and Mr. Roderick, one on each side, kept lancing at it until it was dead. It was dark before we got the fluke chains on.

Tuesday, October 17, 1882

Fine weather with light trades but nothing doing. Some of the boys are making small ivory crosses and some are sewing. Three days ago the captain asked me if I could make a drawing of a paperweight that he wanted to make out of ivory. He described one of a woman lying down on her elbow that he had seen on the ship *Milton*. I got the idea and drew one for him on the leaf of my nautical almanac. Today I showed it to him. The captain seemed highly pleased with it. That was what he wanted, something to go by. He will have some work cutting it out of ivory, but Mr. Gifford says that he is a fine carver and that I should see the set of fancy chessmen that he had carved. Mr. Gifford wanted to know where I learned to draw and I told him that I had never learned, just picked it up.

ST. HELENA

Wednesday, October 18, 1882

We sighted St. Helena today and made for Jamestown harbor. A boat was taken down from overhead for a shore boat. The captain told me to put him on shore and to bring off the mail.

I returned to the ship as soon as I got it. There was quite a package of mail for all hands. I received two letters. After reading them, I laid down listening to the music on Ladder Hill floating over this quiet and peaceful place.

Thursday, October 19, 1882

Some of the men fishing alongside today caught some fine large Spanish mackerel. The rest of us were busy getting ready to get the oil out to send home. There were tackles to get up over the hatches and the cooper was getting ready for a busy time. Mr. Roderick and Sam Hazzard were going on shore and although I did not have time to clean up, I went with them.

Just as I was going up the jetty, Mr. Jamieson slapped me on the shoulder and said, "Come on up to the house, Bob," and would take no excuse. I could not have been more welcome anywhere. The family gathered around me shaking hands. I stayed until after tea, telling them all about humpback whaling and the black people. I had a right royal Scotch evening until the gun boomed. One of the Jamieson girls went down to the gate with me.

Saturday, October 21, 1882

Yesterday and today we broke out oil to ship back home, helped the cooper tighten the casks, and towed them alongside the *Lottie Beard*. Mr. Gifford sent Mr. Roderick and me over to help stow them in the hold of the schooner. When I got back, I shaved and dressed, for our watch was going on shore until Sunday night.

When I got there, I went up to the house at the end of the street and used the knocker. Lady Ross herself opened the door. The pleasant countenance of the old lady did me good to see and I knew that I was welcome. Right away she asked, "When did your ship get in and are you going to stay all night?" She called one of the servants to get tea ready. I had to tell her all about what I had been doing on the west coast of Africa. It came midnight before I knew it. I went off to bed thinking how the men at the ship believe that I sleep at the barracks and I let them think so. This dear old lady knows how to keep me on the straight and narrow path.

Sunday, October 22, 1882

At ten Lady Ross and I walked the whole length of town to the Episcopal church, the English church, as she calls it. We went into a pew well up in front and saw the Jamieson family not far away. The church was very well filled, but there were very few sailors and only three whaling captains. When we came out, the Jamiesons were waiting and we went home to dinner with them. One of the girls came back with us up to Lady Ross' house. After tea the old lady, being tired, rested, while Jennie and I walked to the Governor's garden, where we sat and talked. Then we went on to the promenade and looked at the old whalers so quietly riding at anchor. She told me how she had missed me when I sailed from here the last time. All too soon, the gun boomed and its echo resounded from rock to rock. She held on to my arm until the very last minute at the gate where my friend MacGregor was on guard. I got on board, read a chapter in the Good Book and turned in to my bunk.

Tuesday, October 24, 1882

Yesterday we got out about two hundred barrels of oil and on to the schooner which is going to take it to New Bedford. Today we were at it again until four o'clock, when we knocked off. I cleaned up and went on shore up to the barracks among the Scotch soldiers. We had a good time talking. They wanted me to tell them about whaling and about the Congo, and invited me up to a dance on Saturday night. I gave the Captain a coconut dipper that I had made and he prized it very much. He said, "When I write home, I will tell them about this dipper which was made by a Highland boy on the coast of Africa." It seems no matter where I go, I am sure to meet a good honest Scot, a man that I am not ashamed to grasp by the hand. At the bugle call, all lights went out and we turned in.

AFRICA, TO ST. HELENA

Wednesday, October 25, 1882

I awoke with the thrilling bugle notes ringing in my ears and echoing from the hills. We had breakfast of oatmeal porridge, goat's milk, eggs and bacon. Before our meal, Second Lieutenant MacDonald said a Gaelic grace. I went down Ladder Hill to the jetty and got in the boat waiting for our men. Changing into my overalls when I got on board, I went down in the hold among the greasy casks helping Sam Hazzard straighten out. It's always in a mess after getting the oil out. We worked hard all day, and after filling the deck tub had a good scrub and bath.

Sam asked me if I had a good time on shore. He and the other men always think I go up to the barracks when I go on shore at night and I do not make them any the wiser. Sam has a girl here, like many more, and has asked me up to her house, but I prefer to be under the care of the godly old lady who has been so good and kind to me, if for no other reason than my love of the mother tongue. I like the way she keeps me straight, caring for me just as a mother would.

Thursday, October 26, 1882

We got all of the oil out and on board the schooner *Lottie Beard* late this afternoon. It was a hard job. Sam and I helped Mr. Roderick sheet the large casks in place, well chocked up with the bilge free and made a good job of it. Jim Lumbrie kept tally for Mr. Gifford, one hundred barrels of sperm oil and 477 barrels of humpback oil. We have our after hold all ready for fresh water. I cleaned up and went up to the Jamiesons where we had some good old Scotch songs and some music and talked about Paisley and the places around it.

Saturday, October 28, 1882

Yesterday we took on two hundred barrels of fresh water and ran it down to the casks in the lower hold. The cooper was busy setting up casks to take the place of those we sent home. Today we got seven tons of wood and stored it in the fore hold. At eight o'clock I went on shore and talked to Jennie Jamieson for a while. I had to leave her as I had promised the soldiers to go up to the barracks. When I went into the store room where they held their dance, it was all decorated with paper lanterns and had a platform at one end for the pipers and the musicians. The sets were forming so MacGregor, MacDonald, MacNeal, and I, and four girls made up one. The girls were not so bad but they smelled of whiskey. The one who was dancing with MacNeal was quite tipsy. There was a sword dance after the lancers, then another Scotch dance and some square dances. Two lads did the Highland fling with the pipes playing for all they were worth. I had a glorious night and turned in at the officers' quarters.

Sunday, October 29, 1882

I awoke to the notes of the bugle calling us out of bed. The pipers were marching up and down in front of headquarters. The soldiers were forming and answering the roll call after which they were dismissed for breakfast and to get ready to march down for the church service. When I went down the hill, I met Mr. Jamieson who wanted to know if I was going to church today. I told him that I could not go as I had promised to go on board the ship and probably would not be on shore again before Tuesday.

Tuesday, October 31, 1882

Six men were painting the ship outside and my job was lettering the stern. When I finished and cleaned up, I went on shore and straight to Lady Ross' house. She knew that I was up at the dance on Saturday and asked why I did not come up to her house to sleep. I told her that it was not over until late and that I did not wish to disturb her. She said that it was never too late to open the door to a Highland man. She talked about the coconut dipper I had made for her and said that she was very proud of it. I told her about my adventures on the coast of Africa and all about getting lost in the jungle and the big black men and how they live. Lady Ross said that I was like a son to her although she had never had a child of her own, and that she always felt glad when I came up to her house. She had me kneel beside her and listen to a long Gaelic prayer. What a prayer that was! All for me and so earnest.

Thursday, November 2, 1882

The men all got on board this morning in time for work. Tommy Wilson, one of the harpooners, got his discharge. The mate was not satisfied with him and let him go, for he wants a big, burly fellow. In his place we have another boatsteerer by the name of Grinnell, a big, good-natured chap who weighs about two hundred and thirty pounds and who was third mate on the *Mattapoisett* before we got him. The cabin boy who has been on shore for a week, hiding, was brought on board in irons.

When I went on shore today I met Jennie Jamieson and took a walk with her to Lady Ross' to say good-bye. The old lady gave me some good advice and told me not to forget to read the Good Book. Taking my arm, the girl and I walked down to the gate. I hated to say good-bye to these good friends.

Sunday, November 5, 1882

Yesterday we set sail and up anchor and lay off and on across the harbor until the captain came on board. At noon we headed for Tristan da Cunha. Today, after sighting a school of porpoise, Mr. Gifford went out on the bow of the boat on the martingale and struck one. We soon hauled it on deck, cut it up and had fresh meat for supper. When I came down from the lookout, I had a talk with Grinnell who is a New Bedford man and good company. He is the third harpooner that Mr. Gifford has had this voyage. It seems that the mate is hard to please, but the rest of us have no trouble. I got along with all three of the officers I have steered, Mr. McKenzie, Mr. Reed, and also Mr. Roderick, who is fond of me.

Friday, November 10, 1882

The past few days we have put chafing gear on the rigging and put all our boat gear in good shape. We got out smaller sails as we are expecting stormy weather to the south. The old mast coatings were taken off and new ones sewed and tacked around the fore, main, and mizzen masts after which they were given a heavy coat of paint to keep them tight and to prevent them from getting rotten. Heavy storm sails were bent on for rough weather. We rove new reef tackles on the fore and main topsails, and heavy sheets and tacks on the foresail and mainsail. There are quite a few "gonies" around the ship these days, that is the sailor's name for albatross. Today there is not much doing. I have been making a small ivory anchor for a watch charm and some very small ones with gold catches for earrings for my sister. I have been working on them for the last month back. Now I have them all polished up with shark skin and they look fine and very neat.

Saturday, November 11, 1882

Tonight about ten o'clock, a gale of wind blowing hard, all hands had to be called to shorten sail. With the large crew that we carry, it only takes about ten minutes to strip her of all sail. We are getting down where we can look for squally weather at any time.

TRISTAN DA CUNHA

Sunday, November 26, 1882

Outside of seeing plenty of finbacks, a school of bottle-nosed grampus, a sulphur-bottom whale or two, there has been nothing doing. We have had four or five days of heavy gales and rough seas. Now it's fine weather again and we sighted Tristan da Cunha. The wind hauled to westward and varied considerable around this group of islands. These islands are high and rocky, very steep and forbidding, yet I would like to get on shore to look around and have a talk with the hardy islanders. It is a dangerous shore to be caught on with a bad storm from the southwest. While I was teaching John Brown and Frank Gomez their lessons today, they asked lots of questions about the people living here.

Wednesday, November 29, 1882

This day came in with a gale of wind from the southward.

From the way it felt at masthead, the wind must have been blowing from off the icebergs at the Horn as it was ice cold. We hove to under the close-reefed main topsail and storm trysail. There is a nasty choppy sea and once in a while a snow flurry.

Friday, December 1, 1882

This morning while we were breaking out some provisions, the melodious voice of Grinnell rang out, "There she blows, blows again and again, a school of sperm whales!" We lowered away all four boats and got a good chance to sail right into the school. Mr. King got fast, then Mr. Gifford, then Mr. Hazzard, and finally I got fast. Then the fun started. Whales were running in all directions and got the lines all tangled up. The old mate kept yelling, "Don't cut any line until you have to." Mr. Gifford's whale was close to our boat and Mr. Roderick got a good chance to lance it. I sent our boat on top of our whale, lying there snapping its jaws and lashing its tail all around. Sam Hazzard's whale towed him all around on the outside of the mess. Sam kept shouting to his little red-headed boatsteerer to get close to his whale so he could lance it. After we got our whale killed and alongside of the ship, the captain sent us back to help clear up some of the mess. We got there just in time to help Mr. King whose line was all foul. We put an iron into his whale and the mate told Mr. King to cut his lines as his boat was almost under and half full of water. We had his whale fast and killed it. We were lucky in getting all four whales to the ship.

Saturday, December 2, 1882

Fine weather today and no wind but a big swell. We had the whales all cut in by noon, for the men, all feeling good, worked with a will. After dinner, all four heads were cleared away and the cases bailed out. The blubber was taken below in the blubber room. When supper was over, we started the pots boiling oil. It was my first watch. Mr. Roderick and I were boiling away and firing up with the ship rolling in the heavy swell.. There was no wind to keep the ship steady. We had to be careful not to let the hot oil slop over and catch fire. Mr. Roderick and I were talking about the tangle of the boats in the middle of the school. Three of the men complain of having the mumps and are all swelled up and do not feel so good.

Tuesday, December 12, 1882

This day came in with rough weather and we are cruising along with close-reefed topsails. Sam Hazzard spied a whale and sung out. All four boats went after it but the whale got to going to windward so fast that we could not catch up to it and it got out of sight in the rough sea. The captain hung the colors at the mizzen peak and called the boats on board. In the afternoon Mr. King sighted a lone whale, and we gave chase with all four boats. Mr. Gifford ran right on top of the whale and nearly had his boat upset. However, he got fast. This whale was killed with a dart gun and so we had no trouble with it in the rough water.

Wednesday, December 13, 1882

We had considerable trouble cutting in yesterday's whale on account of the ship rolling and swinging the blanket pieces fore and aft and all around. The cross-deck tackles could not hold them steady, the ropes were that slippery and greasy. It is a hard and dangerous job with the blubber swinging around and across deck, to watch your chance to lower it hurriedly through the main hatch into the blubber room. The seas are heavy so boiling oil has to be done very cautiously and that is slow work. It won't do to let the pots slop over. That might start a bad fire.

Sunday, December 24, 1882

All day yesterday we were laying hove-to under storm sails. Last night all hatches were battened down because of green seas slopping over the rail and flooding the decks fore and aft. There was no lookout at masthead yesterday because of the nasty sea. Today, on account of the ship rocking like a cradle, we set the fore and main topsails and dropped the foresail to keep her steady. Mr. Roderick went to masthead and soon called out, "There she blows!" On account of the storm all of the line tubs had been taken out of the whale boats and fastened to the ship's rail. To prevent the boats filling with water, they were covered with tight-fitting covers of waterproof canvas. Before the boats could start, they had to be lowered level with the ship's rail and the line tubs put in. We chased the whale in the high seas, but losing sight of it, had to go back to the ship again. The whale was seen from masthead again, and all four boats were once more lowered. We had not been down twenty minutes when I saw the whale on top of a wave. Mr. Roderick said, "Bob, watch yourself and get a good brace. I am going to put you on top of it in spite of rough water." Just then the bow of our boat bumped the whale nearly sending us sprawling. I drove the iron in good and solid and we sailed right over the top of the whale, the only way to get clear of it in rough water. The whale sounded and came up not far away and while we could not see it we could hear it snorting near by. We all hauled on the line and soon landed on its back. The whale got a crack at one of the oars and broke it and kept milling around until just before dark, when it finally rolled over on its side after an unusually hard struggle.

Monday, December 25, 1882

Christmas day. Called all hands at five o'clock. Although the sea was rough and sail had to be put on to steady the ship, we had yesterday's whale all cut in by ten o'clock. This was a pretty good whale and we got sixty barrels of oil from it. We took only thirty minutes for dinner.

This was a great day for a Christmas but all the men feel happy. I did not think much about it until I came below to right up after supper. For turkey, we had salt horse, fat pork, and hard-tack. Nobody was dull, all happy and merry, laughing and joking. Grinnell said, "Bob, hand me that leg of turkey." It was a lump of salt beef weighing about two pounds. How that big fellow can get on the outside of a big meal!

Thursday, January 4, 1883

A whale we got last Friday gave us sixty-five barrels of oil and one thousand pounds of whalebone. It is some job scraping the whalebone to get it clean, getting the gum off the slabs and then bundling it.

This afternoon we were laying to and at three o'clock wore ship and close-reefed fore and main topsails. I saw a barque bound southeast under short canvas, diving into the seas, throwing the spray clean over the fore yard, burying her lee cathead and her lee rail to the fore swifter and sheerpole when she lurched. She certainly made some dives into the big seas amid a smother of foam.

Sunday, January 7, 1883

This day came in with a light gale of wind and a rough sea. The ship is under double-reefed topsails. Fernand, on the lookout, sang out, "There she blows!" We got all four boats safely clear of the ship and gave chase. Every time we got on top of a big wave, we had to look sharp all around to get a glimpse of the whale for we could not see far from the boat. The larboard, or mate's boat, got fast and got stove, leaking badly. We pulled up to it and the mate got into our boat. He put another iron into the whale from our boat. The mate told me to get into his boat, cut the line and take it back to the ship out of the way of the others. The three boats working around the whale in the rough water soon killed it. We worked the ship alongside the whale and got the fluke chains on. With the ship rolling heavily when we were trying to hoist the head, the two large heavy chains parted and the head rolled under the ship and was lost. Without the head, this whale gave us sixty-eight barrels of oil.

Thursday, January 18, 1883

Down in this latitude, about 47°S. it is kind of cool on deck. There's not much to do except watch the "gonies" coming alongside, looking up at you as if asking for something. Sam Hazzard got a hard-tack and broke it in small pieces. These big birds flew up and took the pieces right out of our hands. A flock of Mother Carey's chickens kept following the ship all day.

Sunday, January 21, 1883

In the afternoon all hands were on deck enjoying the bright sunshine. John Brown was sick in his bunk but I gave Frank Gomez his lesson. He manages to read fairly well now and knows the numbers. When I first got hold of him he knew nothing about figures and could not tell time by the clock, for clocks were few and far between on the island of Flores. Frank was telling me of a family whose son brought a clock home and then went to sea again. None of the family could tell the time but they would wind the clock every day. When they wanted to know the time, they had to get a man who had been to sea to come and tell them. Time never bothered them. Daylight was morning, noon a guess, and night was sunset. That was enough for them all.

Sunday, January 28, 1883

Today we are in Lat. 26° 31' S. Long. 9° 52' E. with very pleasant weather, light southeast wind and all sail set. A large four-masted ship went by close to us, with a cloud of canvas. She carried double topgallant sails on the four masts, royals and skysails, and looked like a Dundee clipper. What a fine sight she was with a beautiful figurehead and her bright spars shining in the sun and her enormous spread of snow-white duck! It does the heart of a sailor good even to look with pride and admiration as she goes by standing up like a church steeple. There were two men at the double wheel, steering, leaving a straight wake astern. She was steady as a rock and curled a feathery foam at her bow.

Thursday, February 1, 1883

The last few days we have been having pleasant weather with light trades. The men have been working on the rigging and patching sails. We unbent the flying jib, re-roped it, stitched the middle seam, put a band across from the clew, and bent and set it on the stay. The heavy mizzen storm staysail was taken down and an old patched one put up in its place. While I sat on the main hatch, sewing, one man threaded my needles and waxed the twine. We examined and overhauled the footropes and braced the pennants, parceling and serving where worn. The pennants, footropes, and stirrups were tarred hot. The good foresail was unbent and stowed down for future use. Up went the old patched one, which looked like a crazy quilt but good enough for this mild weather. There are very few storms here where we are cruising now, that's why we put up the patched sails and keep the good ones for the stormy cruising grounds to the south, where the wind and waves are both strong and high.

Sunday, February 4, 1883

Having nothing to do, I took an observation with my old quadrant and made out Lat. 19° 12′ S. Long. 9° 32′ E. I held school for my two scholars, Frank and John, and had them writing on a slate that I have with me. I write on it when I can't find time to write in my diary, so that I do not forget important things. Mr. Gifford said, "Let me see that slate." He showed it to the captain who said that the writing was good.

Monday, February 5, 1883

Overhauled the fore stays and the martingale. We made new footropes for the jib boom and bowsprit. The headgear and sprit guys were overhauled and tarred. Some of the men were working on the topgallant rigging, footropes, and lifts, taking down the old ones and putting on new ones after tarring them well.

We changed ends with clew lines and buntlines, also the topsail halyards. The halyard blocks were overhauled and blackleaded. Birds were sighted around a waterlogged plank so we hauled aback and lowered a boat. The plank was covered with barnacles and had lots of fish around it, mostly sea bass, quite a few of which we caught. We got the plank on deck and scraped off the barnacles. It was of white pine, a board three inches thick, fifteen inches wide, and twenty feet long.

Sunday, February 11, 1883

All sail set in good weather, but hazy. All of the officers and harpooners are gathered round in midship. Grinnell is making a model of a ship, or rather a half model. I am helping him cut out the wooden sails. He is a big good-natured fellow yet he is quick and light on his feet and can dance a hornpipe fine although he weighs two hundred and thirty-five pounds. I am reading a book on navigation that the captain loaned me. Mr. Roderick asked if I did not get tired reading but I told him that I would read and study much more if I had the books and the time. He went below and brought up a lot of New Bedford papers and asked me to read all the whaling news to him. Mr. King and the cooper with some of the men crowded around listening to the report of the catch of each vessel. The men enjoyed the reading of the reports of the ships in from the Arctic Ocean and of ships that they knew. Mr. Roderick asked over and over again about the forty whalers that had got back home.

Wednesday, February 21, 1883

There has not been much doing lately, lots of cruising but no whales. I have been studying navigation the best I could without a teacher. Grinnell has his model of half a ship done. It looks neat, is well made, and will make quite an ornament. Some of the men have quite a fad of making small ivory crosses, others are making chains of tortoise shell and ivory handles for canes. Captain Howland finished his paper weight, that I drew the design for, a woman reclining with one elbow on an urn. He asked me how it looked. He did a fine piece of work, beautifully polished. The base took a whole sperm whale's tooth and the figure another. He fastened them together underneath so no one could see the pins.

Thursday, March 1, 1883

At daylight we set all light sails, for it is fine weather and there is a good breeze. When I came down from masthead at two in the afternoon, after seeing nothing but my little friends the tropic birds, the captain said, "Bob, take a sight for me because Mr. Gifford cannot as he has something in his eye." He let me use his sextant. I took an observation and gave him the altitude. He gave me the time by chronometer and I worked out our position on the slate and left it on Mr. Gifford's desk. The mate's eye being very sore, he asked me to put the observation Lat. 15° 55′ S. Long. 4° 39′ W. in the log. Sam Hazzard asked where we were. I told him that we were about sixty miles east of St. Helena. Just then the bell rang for dog watch, four o'clock. At the same time Mr. Roderick sang out from aloft, "There she blows, again and again, one lone whale!" The orders came quick and fast, "Get the boats ready. Haul aback the main yard; hoist and swing all four boats; shove clear of the ship." We up mast and sailed for a good distance. The whale came up not far away and was heading toward us, so we took it head on. Mr. Roderick gave me a good chance with no fear of bungling. I drove both irons in up to the sockets. Down went the whale taking nearly all the line from the tub. I was busy getting the mast and sail down. Mr. Roderick was shouting to keep clear of the line which was a-humming, and told me to pass him the mast and sail aft. After we exchanged ends, up came the whale with a loud snort and started to speed away to the south like an express train. In fifteen minutes we were out of sight of the other boats. Again the whale sounded but not so deep. When on top of the water, it kept going at an awful speed. I was at the steering oar and Mr. Roderick was standing in the bow with his lance waiting until we got close enough to lance it and calling, "Haul line, haul line everybody and get up to the whale." At dusk, the whale getting tired, we got close but it turned head on toward us, rolled over and tried to chew up the boat. Mr. Roderick got a good chance to shoot a bomb into it. I did not need to tell my boys to stern all for they had seen those enormous jaws with those two rows of teeth ready to come down on the boat. The danger was soon over and the whale dead. We pulled up close to it, coiled down the line and straightened out the things in the boat. Looking around the waste of water, no other boat was to be seen, not even the ship. We opened the lantern keg, got out the lantern, lit it and hoisted it up on an oar. Mr. Roderick gave out the twelve hard-tack that were in the keg and pipes and tobacco to the men, except to me, for ·I do not use tobacco. There was a box of matches and six candles for the lantern. The men drank all the water and ate all the hard-tack. We slacked line and kept back from the whale. Having hardly any clothes on, the men lay down close together to keep warm. You see when we go down in the boats in warm weather, all we have on is a pair of short dungarees or short canvas pants and an undershirt so that if the boat is knocked to pieces we can swim around easily and get hold of an oar. While the men lay down huddled close together, Mr. Roderick and I kept watch until morning.

Friday, March 2, 1883

When the sun came up, there was nothing in sight. The sun warmed us and it felt good. The day seemed very long, especially as we had nothing to eat or drink. The only thing in sight was the dead whale. The trade winds blew strong all day. In the afternoon the sun was broiling hot, making us very thirsty. Now and then a man would get up and look around the horizon but there was not a thing to be seen. These men did not realize what they were up against. All of the men in the boat, except myself, were Portuguese. When night came again and no boat nor the ship had been seen, they began to get restless, so I had to talk to them and encourage them. After dark it got cool, and the men huddled down together in the bottom of the boat. Taking the boat sail, I covered them up. They were soon sound asleep. After all was quiet, I went forward and took the knives and hatchets and put them aft under the den.

Saturday, March 3, 1883

Today the whale was beginning to smell. To try to stave off our hunger with blubber or meat from the whale was sickening. Mr. Roderick kept saying, "Look, look again," but we saw nothing. The men were getting quite restless, hungry and parched with thirst, so both Mr. Roderick and I talked to them. Even Mr. Roderick himself was getting uneasy and asked what we had better do, whether we should cut loose from the whale or stay here. I said that we had just as much chance to be picked up here as any place else. He asked if we could run for St. Helena. I told him that if I knew where we were we might make a try for it, but I did not know which way to steer. St. Helena lay to the north and west but how much was the question. The island could easily be missed and then we would be headed for the middle of the Atlantic Ocean. Tonight Mr. Roderick kept first watch and woke me about midnight. I fell asleep myself near daylight.

Sunday, March 4, 1883

Mr. Roderick asked me what time it was so I looked at the compass and got a bearing on the sun and told him it was about ten o'clock. He wanted to know how I could tell by the compass. I tried to show him but could not make him understand. I told him that when we struck the whale we were sixty miles east of St. Helena and about twelve hundred miles from the coast of Africa. He wanted to know how I knew. I said, "You saw me taking the sun for Mr. Gifford and I worked out our position."

It was awful hot in the afternoon and when I lay down, I could not go to sleep on account of the awful thoughts running through my head. I was thinking of the story that George Pollard told Captain Fisher and me when in the cabin of the *Abbie Bradford* about the crew of the *Essex*.* At night when

* In November, 1819, the Nantucket ship *Essex* was cruising in the Pacific whaling grounds. The chief mate's whaleboat was stove but reached the ship. Meanwhile, the captain's and second mate's boats were fast to a whale. The ship headed toward them. Suddenly an eighty-five foot sperm whale

I lay down in the stern sheets, I prayed as I never prayed before, asking that I could maintain my mind and guidance through this awful trial.

Monday, March 5, 1883

This morning the wind was blowing strong and the sea was getting rough. Mr. Roderick looked frightened. I looked at the compass and found the sun bearing north at noon. I told Mr. Roderick that the way the whale was bumping into the boat might cause it to leak, so we stuck a waif in the whale and cut loose. We put up the mast and reefed the sail.

Mr. Roderick said, "Bob, how about you taking charge?" "All right," I said. "We will leave the whale and leave for the coast of Africa, for the ship may never find us. With good trade winds, we should get there in eight or ten days, maybe sooner, or we might meet some ships from around the Cape. We can make 150 miles every twenty-four hours, but we will have to talk to the men to keep their courage up." I also told Mr. Roderick that we would have to do our best to look cheerful. I shook the reef out of the sail. With the men all on the weather gunwale, the boat bounded along throwing spray until we were

---

breached only a hundred yards away and headed full speed for the *Essex*, striking her forward. Once again the whale rammed the ship, staving in her heavy planks close to the catheads. In two minutes the ship was on her beam ends but did not go to pieces until the third day. In the meantime the men repaired their boats and provisioned them. On the third day the ship was abandoned and the boats headed on their long row to the coast of Peru, three thousand miles away. Five days later, they reached Ducie's Island which was barren and without water. Yet three men preferred to stay there rather than endure the tortures of heat, thirst, and hunger in the open boats. One by one the men in the boats succumbed to thirst, hunger, and exposure. As they died, their companions fell upon their bodies and devoured the raw flesh like famished wolves. Lots were even drawn to see who would be killed to save the others.

The boats became separated. On February 17, 1820, the chief mate's boat with three survivors was picked up by the British boat *Indian*. Five days later the ship *Dauphin* of Nantucket picked up the captain's boat with only the captain and one man left, three months after their ship had been rammed. Such is human endurance.

all soaking wet. The weather was awful hot but the spray helped to take the drouth out of us. The men are plucky but the lack of food and water is beginning to tell on them. I went to sleep and awoke with the mate falling on me when the boat gave a lurch.

Tuesday, March 6, 1883

At noon or thereabouts, the whaleboat going along at a good clip with all the sail that she could carry, we saw smoke on the horizon and ran for it, but it faded away and we could see nothing. We were thankful for a good breeze that boomed us along. Every once in a while one of us stood up and looked all around but there was nothing to be seen but the waste of water. While steering this afternoon, I was wondering if we could hold out without any serious trouble until we reached the coast. Our old boat was footing right along and the breeze was good. Some awful thoughts ran through my head yet I had no fear. The men are still very good, but when hard pressed and tried too much it is hard to tell what they might do. I was awakened in the night by a jabbering in Portuguese. The men had seen a light once in a while as the boat rose and fell but when we tried to run for it, it was too far away and we lost sight of it. That made the men feel low.

Wednesday, March 7, 1883

I awoke sometime in the forenoon and took the tiller. I told the mate that I thought that we were in or near the track of ships coming around the Cape of Good Hope and that it would be wise for the men to take turns standing up for a short time and look around the horizon not only during the day but also a lookout at night. The men seem to be getting weaker because they are sleepy and drowsy all the time and keep lying in the bottom of the boat. They awakened me in the middle of the night. They had sighted a sail in the moonlight, but it was too far to row in their weakened condition. It is scorching hot in the daytime but it is chilly at night.

Thursday, March 8, 1883

This day the wind was light. We saw nothing. The men say nothing but look wistful and sad. Their lips are dry and cracked. At night as it got dark, I dreaded going to sleep. I told Mr. Roderick that I would take the first watch tonight as I was not sleepy. It is just a week today that the whale took us.

Friday, March 9, 1883

I awoke Mr. Roderick at daybreak. As I did not sleep all night I was very drowsy and slept until late in the afternoon and felt refreshed, but my mouth was parched and my lips were split. The sun is awful. I did feel queer and kept thinking of things to eat and drink but I knew that I had to rouse myself, so I stood up on the covering board and looked around but saw nothing. Late in the afternoon, five or six albatross came flying around. This discouraged the men as they are very superstitious.

Saturday, March 10, 1883

Today the wind was light. Mr. Roderick looked bad. I told him not to look so gloomy as that would discourage the men but he said he could not help it. He told me that he kept thinking of his wife in New Bedford all day and could not get her out of his mind.

Sunday, March 11, 1883

Things look bad in this boat. I can't get the story of the *Essex* out of my mind. Mr. Roderick is very melancholy tonight and said, "Bob, let me sleep so that I can forget." I stayed all night at the tiller.

Monday, March 12, 1883

I awoke with Mr. Roderick calling me. He said, "Bob, I thought you were dead." He had been calling me for some time. The tears were running down his cheeks. It hurt me to see this big strong man in tears and made me feel bad. I had a long talk with him this afternoon and he felt better. At night I had him kneel down and pray alongside of me. He had his beads around his neck and counted them over. It is his belief and he is sincere. When he lay down to go to sleep, he said good-bye as tender as a woman and as if it was his last night.

Tuesday, March 13, 1883

I awoke sometime in the forenoon with the boat lying idle. The mate was sound asleep and the sail was flopping loose. I took the tiller, hauled aft the sheet and got headway on. I did not think to look around. All of a sudden, I saw a large barque heading straight down on us, about two miles to the windward. They rounded to, hauled aback and took us on board. All of the men were given a drink of brandy. I took a cup of coffee and a little to eat. I cautioned the men to eat very little at first or they would be sick. The captain called Mr. Roderick and me aft, and asked how we came to be in such a predicament. We told him all about it. He said that he would take us to St. Helena and land us because he was going there anyhow for water. This ship was the barque *Fearless* of Shields, England. The chief mate was a Scotchman from Stornaway. They treated us fine. The captain had the mate and me eat in his cabin. This was one fine big ship. After supper I fell asleep, feeling thankful that we had good firm planking under our feet once more.

Wednesday, March 14, 1883

Today I looked over the ship with the mate, who although a Highlander does not speak Gaelic. The ship has a fine crew of English sailors. In the afternoon Captain Nottingham, an Englishman and a good sort, called me and wanted to know how our second mate could be such and not be a scholar nor be able to read and write. I told the captain that it was his whaling qualifications that let us have a mate who could not navigate. I told him that our ship, though small, had four mates and that only two needed to be navigators.

ST. HELENA
Thursday, March 15, 1883

Had breakfast with Captain Nottingham this morning. All of our men from the *Kathleen* are feeling well in spite of lack of food and water for eleven days. I went aloft to the main royal yard and sighted the island of St. Helena. When I came down and reported to the mate, he said that we would be there about three o'clock. The *Fearless* did not anchor, just hauled aback and signaled for a water boat. When it came alongside, I thanked the captain and all of us climbed in. Mr. Roderick said, "Let us go on board of the *Andrew Hicks*." We told them our tale of woe. They asked us to stay all night and told us that the *Kathleen* was expected any day. Their boat was going on shore with the watch. I wanted to go on shore too, but I was barefooted and had no hat, only a pair of pants and a blue shirt. I went anyway. I did not know where to go or what to do. I finally decided to go up Ladder Hill. There I received a Highland welcome and a good supper. The soldiers sat around until taps, listening to the story of our mishaps. I slept sound at the barracks until the sound of the bugle in the morning.

AT ST. HELENA, THENCE TO MAYUMBA, AFRICA, AND BACK TO ST. HELENA

The lost crew returns to the Kathleen—*Transferring oil—Swimming match—Scolding from Lady Ross—Whaleboat enters race with soldiers and sailors—Gala day of the race—Whaleboat wins four sovereigns in gold—Girls—The consul's daughter—Ferguson advanced to fourth mate—Off for whales again—Return for runaways—Nearly run down by ship without lights—Making clothes—Stove boat—Humpback whales again—Shark meat for the natives—Mayumba —Back to St. Helena.*

CHAPTER IX

AT ST. HELENA, THENCE TO MAYUMBA, AFRICA, AND BACK TO ST. HELENA

Saturday, March 17, 1883
I had breakfast in the officers' mess. All of these big Highlanders were very kind to me. One came to me with a pair of brogues and a pair of fancy homespun stockings and told me to put them on. I did not want to take them but he insisted and they just fit. I thanked him until I could better repay him. As we were eating dinner one of the men came and reported two more whalers in the harbor. I went out to look and sure enough, there was the *Kathleen*, coming in to anchor. I thanked the men and officers for their kindness to me, hurried down to the steps of the jetty and got on board the *Andrew Hicks*. The second mate, Mr. Roderick, and the four men were waiting for me. We got alongside the *Kathleen* before the anchor was down and climbed on board. All hands were glad to see us. Captain Howland spoke very little to me and less to Mr. Roderick. I had to tell Mr. Gifford all about how we got along from day to day and how we got here.

Mr. Gifford was awful good to me. He told me that they found our whale two days (as I figured it) after we had sailed away from the carcass for the coast of Africa. It was a very large whale about eighty feet long and gave one hundred barrels of oil. Mr. Gifford said, "The night you got lost, you were out of sight before we had our mast and sail down. As for the ship, she lost sight of you in less than half an hour." I told him that we had killed the whale just at dusk. He told me how they had looked and had cruised back and forth in the direction we had last been seen, and when we were not to be found they did not know what to think. About the sixth day we had been gone, the man at masthead reported a large flock of birds about five miles away. When they ran down, there was the dead whale with the waif and both of the waist boat's irons in it. Then the ship did not know which way to go. The mate asked me what Mr. Roderick asked me to do and if the men got rough. I told him that Mr. Roderick asked me to take charge on account of my being able to steer a course and that our boys behaved like men even if they had nothing to eat or drink after the first night out. I told Mr. Gifford that I figured on reaching the coast of Africa in the strong trade winds and to get in the way of the ships from around the Cape.

Mr. Gifford said, "Bob, you done just what I would have done." He also told me that they got the whale in Lat. 18° 39′ S. Long. 7° 48′ W. which made it 260 miles from where we struck it. We both talked until ten o'clock, for he wanted to get all the particulars to write in the ship's log. He told me that Mr. King was sick and that the captain was thinking of letting him go but to say "nothing to nobody" yet.

Sunday, March 18, 1883

A fine day. Got up, washed, shaved, and put on my good clothes. Mr. Gifford asked if I would like to go on shore. I said I would like to go on shore and to church. "All right," he said, "but come off early tomorrow as we have the oil to ship home." He called some men to put me on shore just as the church bells were ringing.

There was a good sermon. I met the Jamiesons as I came out, and had to go to their house and have dinner. When I got in the house, I got a good scolding from Mrs. Jamieson, for she had heard all about my being lost from the ship. It seems that Mr. Jamieson heard about it from the mate of the *Fearless* when he came on shore. He had looked around for me but I had gone up to the barracks. I told Mrs. Jamieson that when I came on shore I was barefooted, nearly naked, and had no money. To which she said, "All the more reason you should come among your Scotch friends." After dinner I told them about the adventure. They wondered if I had been afraid of the Portuguese. I said that they were all good men who had a liking for me and that I was only worried about them for having no food or water.

Wednesday, March 21, 1883

Last night the mail came on board with three letters for me so I stayed on board to read them. I was glad to hear from

AFRICA, AND BACK TO ST. HELENA

home. Today there was a swimming match across the harbor between a soldier from the Ninety-third Highlanders and a sailor from the gunboat *Sea Gull*. The soldier's name was Cowan and he won the match easily. He came from Glasgow and was a champion swimmer in Scotland. The *Sea Gull* is the gunboat that we saw in the Indian Ocean, now stationed on this coast for a short while as she works her way home.

At four o'clock I went on shore with the first boatload and met one of the soldiers who told me there was going to be a boat race soon. I told him that I would like to get in the race, so he said to come up to the barracks and have a talk with the captain who knew me. He thought it was quite likely that matters could be arranged.

Then I went up to Lady Ross' house. She gave me a scolding and a blessing at the same time. She had heard all about my adventure from the Jamiesons and said that I had a right to come to her house until my ship came in. I felt sorry to have offended this dear old lady, but she soon put me at ease. I told her about the swimming match. She was pleased that the Highlander won. I told her that there was going to be a boat race between the soldiers from the barracks and the sailors from the gunboat and that I wanted to get my whaleboat crew into it, for they are the best crew on the *Kathleen* and have beaten several other ships' crews. Just before bedtime Lady Ross asked me to read a chapter in the big Bible. After that we knelt down, and such a prayer as she said for my deliverance from the vast and mighty waters.

Thursday, March 22, 1883

After breakfast with my dear old friend, I went up to the barracks to see the captain of the Highlanders. The lieutenant of the gunboat *Sea Gull* was there too. When he saw me, he remembered having met me at Johanna in the Indian Ocean. He was glad to see me and came over and shook hands. I said that I had heard about the boat race and that I would like to get into it. He said, "You can't pull against these narrow boats of six and eight oars with only five men." I said I would like very much to get in the race. The captain of the Ninety-third said that we could make it four boats and that he would compete with the six-oared gig. We all agreed to have it on Wednesday at noon, each boat putting up a sovereign as stakes, the winning boat to take all. At first the distance talked of was one mile, but the lieutenant of the gunboat thought his men would have a better chance at two miles. We came to the conclusion of two and one-half miles, right across the harbor from east to west, and all agreed to it. The soldiers have a fine boat that belongs to the barracks. It has eight oars, a regular yacht's cutter which they use for exercise. The gunboat has a long narrow gig, also a small cutter which is very good. The captain asked me to have a bite of supper with him and the rest of the officers. He told them all about my getting lost from the ship. The lieutenant of the gunboat asked about Sam Hazzard and Tommy Wilson, and we had a pleasant time until the gun boomed for all to get on board.

Saturday, March 24, 1883

Yesterday and today, all hands were busy getting the oil out for shipment on the schooner *Lottie Beard*. Mr. Gifford attended to the hoisting out on deck. The cooper was busy becketing the casks for towing over to the schooner. The second mate was over there stowing casks in the hold. Our deck was all lumbered up with shooks and oil casks. After getting three tons of coal from the barque *Roscoe* of New Bedford, we stowed it down in the forehold.

Sunday, March 25, 1883

On my way to church I watched the sailors from the English gunboat forming in line at the jetty and marching to the church. The soldiers from the barracks marched down hill and followed them with bagpipes playing. The church was well filled with the red coats and kilts of the soldiers and the white duck trousers and blue flannel shirts and straw hats with blue ribbons of the sailors. Although there was a fleet of thirty-five or forty whalers in the harbor, I do not believe there was a man of them besides myself who came to church. After service Mr. Jamieson had me go to dinner with him.

Tuesday, March 27, 1883

Today the water supply boat came along and we ran one hundred barrels of fresh water down in the ground tier of casks in the after hold. After telling Mr. Gifford about the boat race, I asked him if I could take my old boat, the bow boat. He said, "Yes, but you had better ask Mr. King about it." Mr. King said that I was welcome to it. The mate asked why I wanted the bow boat. I replied, "Because it is two feet shorter, eight inches narrower, and lighter than any of the other boats on the ship. My crew was not beaten by any other crew on the ship when I used that boat." The mate asked if I thought I could beat those narrow eight-oared boats. I said, "Mr. Gifford, their oars are only eight or ten feet long while ours are from fifteen to seventeen feet long. Our men sit on the far side of the boat so you can see the extra leverage that our men have, while their men sit amidships. Our men have often pulled ten miles in rough weather with never a complaint. Just look at the long pulls we had when humpbacking. Their men won't last, for they are not used to long pulls."

Wednesday, March 28, 1883

There are a lot of whaling barques lying here: *Roscoe, Alice Knowles, Stafford, Greyhound, Petrel, Desdemona, Andrew Hicks, Sea Ranger, Bertha, Morning Star, Art Tucker, Pioneer,* and *Mars*. Besides, there are the whaling ships *Milton, Eliza Adams, Niagara, Jerry Perry, Hercules, Chas. W. Morgan, Mattapoisett, Seanne, Stamboul, Sunbeam, Wave, John Carver, Europa,* and the *Tropic Bird*. Five large merchant ships are also anchored here. By ten o'clock all of the ships were decorated for the race, dressed up with flags and streamers from the truck to the taffrail.

I called my crew, lowered the bow boat level with the rail

and took out everything but five oars, two fifteen feet, two sixteen feet, and one seventeen feet long, every one of them of good ash. The rowlocks were covered with leather and well greased. I took my crew for a short spin, the boat only drawing six inches of water. We did not go fast. I told my boys what to do and what I wanted of them.

There was a stake boat anchored to the east of the harbor, and to the west of the ships two boats were anchored for the judges. I warned my crew not to pull until I gave them the word. They wore light undershirts, dungaree pants, and were bareheaded and barefooted, every man a Portuguese except myself.

We pulled over to the starting boat at our leisure and watched the men from the gunboat pass us. They were dressed in white duck pants, white shirts, blue caps, and looked very neat. They pulled a six-oared boat against our four. Then the soldiers came along, dressed in blue pants, blue shirts, canvas shoes, and Glengarry bonnets cocked on the side.

My crew was feeling good, laughing, joking, and chaffing one another, but when we all backed up and got hold of the ropes, they were as quiet as mice. We got our instructions and were warned not to crowd any boat or we would be ruled out.

The rigging of every ship in the harbor was full of men looking on. The promenade was crowded with men and women watching. Off went the gun and away we went with cheers ringing from the people on shore and on the boats.

I said, "Now, boys, no talking."

The gunboat crew gained on us a little and the soldier's boat got two lengths ahead of us in the first half mile. The sailors were trying hard to catch up to them. I told my men to keep steady for the first mile, at the end of which the sailors' six-oared boat was abreast of the soldiers. I saw that the soldiers would not hold out. The sailors' eight-oared boat was a length ahead of us. My Portuguese were just as cool as when we started and were not laying down to pull hard. Now I said, "Boys, it is time to get down to it, but do not break your oars. Bend your backs and put a little more force into it." The smiles on their faces told me they were pleased.

We shot by the soldiers' eight-oared boat with about a mile to go yet. The sailors' six-oared cutter was a half length ahead so I sung out, "Now, Portuguese, pull like after a sperm and lay down to it." Every stroke I pushed on the stroke oar with my right hand while holding the steering oar in my left. I could feel our boat jump forward at every stroke. As we pulled abreast of the ships, all three boats were astern of us, the nearest three boat-lengths away.

When we got to the stake boat, my men were pulling just like clockwork, fresh and good for ten miles more. The nearest boat was five lengths astern. A mighty cheer went up from all the whalers. They cheered and cheered, "*Kathleen! Kathleen!*" until the rocks echoed. We were surrounded by all the boats in the harbor. The lieutenant of the gunboat asked me to come on board, but no, he did not ask my crew. I did not tell him *that* when he came on board the *Kathleen* to talk to Mr. Gifford. He asked the mate to bring me along on board his gunboat for dinner. We had a good time and a grand dinner. The lieutenant told Mr. Gifford that my crew looked fresh at the finish while his own crew looked worked out. Mr. Gifford told him it was nothing for my crew to pull twenty miles to windward in a rough sea with never a whimper out of them, nor would they say "tired" once. I got my four sovereigns in gold. When we got back to the *Kathleen*, Mr. Gifford said, "Bob, I feel proud of you. You certainly have that crew well trained." We sat on deck talking until time to turn in.

Thursday, March 29, 1883

My boat's crew helped me put all the things back in the boat that we used in the race yesterday and hoist it on the cranes. I thanked them for the way they behaved in the race and gave Fernand two of the gold sovereigns which he was to share equally with the other men. He did not know how to divide it. I told him that it was forty-two shillings altogether or eight shillings and five pence for each man. It made them very happy.

Friday, March 30, 1883

Tonight I dressed and went on shore. As I started up the jetty, MacGregor, MacDonald, and some of the men from the barracks blocked my way and said they were glad a Highlander won the race. Along came Mr. Jamieson who had me go down to his house for tea with his family. Mrs. Jamieson said, "Girls, this is the Scotchman who beat our army and navy." I had a good time and heard some good music, piano and violin. When it was time to go on board, Miss Flora walked down to the gate with me. She asked if I was not afraid when I was lost from the ship. I told her that I was worried and hoped that I could keep my reason. It was a bitter experience and we might not have been able to hold out many more days.

Saturday, March 31, 1883

Four hundred and sixty barrels of oil all told and twenty-five hundred pounds of whalebone, bundled and cleaned, were put on board the *Lottie Beard* to be sent home this trip. While I was doing a little writing, the captain came on board. When he saw me, he came over and talked about the boat race. He asked how it was that I had managed to do so well. Captain Howland had known nothing about the race until he saw all the flags up on the ships in the harbor. He looked over the wall just in time to see our boat coming in first. Until he heard the cheering for the whaler *Kathleen* he did not know it was my crew that won. He said, "Bob, you have a good boat's crew. They are a credit to the ship. Now, Bob, will you put me on shore?" I replied, "And with the champion crew, sir." He laughed.

Sunday, April 1, 1883

Last night I stayed with my soldier friends at the barracks. At breakfast this morning, the captain asked what time I got up here last night. When I answered, he said, "I bet he was down in the garden with some girl, holding hands." All the men laughed and roared again. I replied, "That's right, and maybe it was your girl I had. You know that sailors are devils for stealing the soldiers' girls." They laughed and I knew I had them. Then we all marched down to the church.

When service was over, Mrs. Jamieson said, "Ma braw Highland laddie, awa up to the house an ha a bite of dinner wi our folks. You deserve it for defeating the soldiers and sailors." I said, "Yes, with a Portuguese crew, bareheaded and barefooted." She laughed and said, "That's nothing. The Duke of Wellington whipped and drove Napoleon's best generals out of Spain with a handful of barelegged Scotchmen and a few hundred Portuguese." After tea, I went for a walk with Flora Jamieson down to the Governor's garden.

Wednesday, April 4, 1883

Yesterday Grinnell came on board, drunk, so I shoved him downstairs in the steerage, out of sight, before the mate would see him.

Mr. Gifford said I could go on shore tonight. When I went through the gate, there was Flora waiting for me as she said she would be. She waited, although I had told her it was not my night on shore. I had not dressed, for I had to be back on board by nine o'clock. We sat down on a bench in the corner of the rocks and in the shadow of a tree and had a pleasant chat. I asked her what her mother would say if she knew she was out with me. She replied, "Mother knows that I am out with you." When it was time to go, she asked me when I would be on shore again. I told her that I would be sure to come to say good-bye to her and to Lady Ross. She said she would go up there with me and as we parted said, "Robert, God bless you."

Sunday, April 8, 1883

Only Mr. Gifford, the steward, and myself were on board today, but we had some visitors, the American consul, a Mr. Thatcher, who was the captain of a whaler some years ago, two other gentlemen, two ladies, and a young girl about sixteen years old. Mr. Gifford told me to show the young lady over the ship and explain things. Mr. Thatcher said, "Young man, you will have your hands full."

The girl could ask more questions than a five-year-old boy and wanted to see everything. I had to take the cover off the try works, show how the windlass hove in the chain and anchor, how the blubber was minced in the mincing machine, and what the inside of a whaleboat looked like. I called my crew, hoisted and swung the boat level with the rail. Nothing would do but that she should get in the whaleboat. She looked over the lances and harpoons, even taking off the sheaths. I was afraid she would cut herself on the hand. I had to show her how I stood to dart at a whale. Just then the mate called Mr. Thatcher up on deck and said, "Bob has your daughter and is going whaling with her." She started to get out of the boat down to the deck, and laughing said, "Catch me," and jumped into my arms. Going aft to the galley, she bothered the cook for ten or fifteen minutes. Mr. Gifford sent me on shore with them. On the way the girl said, "You are the man who won the boat race. I am glad an American won it." I told her that I was Scotch and that all of the men in my crew were Portuguese. Her father asked me to come up to his house when on shore, but not for me. He keeps a grog shop.

Monday, April 9, 1883

The third mate was put on shore, very sick. He got his discharge and is going to a hospital. The watch went on shore for the last time, as we are going to sea again. The captain came off and had some talk with the mate. Then he called me and told me to move into the stateroom with Mr. Hazzard. He said that I was to act as fourth mate and to head the starboard boat. Next, he called the crew to put him on shore and told me to come along and bring a box of tobacco which I was to take to Mr. Thatcher, the American consul. When I went up with the tobacco, Mr. Thatcher asked me to have a drink. I told him that I never drank. I was just leaving when Captain Howland came in. He called me back to sign the ship's papers as fourth mate.

From there I went to Lady Ross' house to bid her good-bye. I told her that the ship was going to sail tomorrow. She wanted me to stay all night, but I could not promise because I should ask permission from the captain first. I shook hands with her and said that if I met the captain, I thought I would be back. I ran up Ladder Hill to say good-bye to the Highlanders in whose company I had spent many a pleasant hour, dancing, playing, shooting at bottles, boxing, and fencing. They were all glad to see me and all wished me good-bye as I started down the hill. On my way to the Jamiesons I met Captain Howland. I told him that I would like to stay on shore tonight. He said, "You've had lots of shore leave this time, but you deserve it. First of all, I want you to take a message out to Mr. Gifford and tell him that I will be up at the hotel."

After delivering the message, Mr. Gifford said, "Bob, the captain says you are always to be trusted. You go on shore but remember I am depending on you to help me in the morning."

When I went on shore again, Sam Hazzard rushed up to me, grasped me by the hand, and with a smile all over his face said, "Bob, I am awful glad to have to address you as Mr. Ferguson, and to have you for my roommate. I only heard it an hour ago when I went on board to beg Bloody Dan [Mr. Gifford] to let me stay on shore tonight."

I went on up to the Jamiesons' and Sam went up Donkey Lane to see a girl. I had a grass mat along with me that I thought Mrs. Jamieson would like. When I spread it on the floor, she laughed and said, "Is this from one of your big black girls in Africa?" I told her how I had traded calico for it with a big naked man in a dugout near the Congo river. After tea, I said that I was going to stop all night with Lady Ross. Flora Jamieson said, "I am going up there with you."

As we walked up the street at dusk, Fernand and Frank saw me and wanted to speak to me. I beckoned to them and asked what they wanted. They had no money, so I gave them each two shillings. I told Fernand to take his things and move into the steerage and to look out for the starboard boat as my boatsteerer. He thanked me and said that he had heard at noon that I was fourth mate. The two boys went away smiling and looking at the girl waiting for me.

Flora asked, "Are they some of your boat crew? Why do you call them boys when they are older looking than you are? That dark complexioned one you call Fernand is a good-looking chap."

I replied, "He is a good and smart lad and is my harpooner now. His people are well to do."

She wondered why I gave the boys money, thinking they would get drunk. I told her that they might drink wine but would never get drunk.

Lady Ross was all smiles when she saw me with Flora. We went in and after talking about Scotland for a while, the girl played on the piano and the old lady sang a very touching Gaelic song, "The Farewell to the Clan," with a great deal of feeling.

As I took the girl home, she said, "You missed my older sister, Jennie. She got married two months ago to a man who came here on a visit and is now living in England. She was very fond of you."

Flora wanted to know when my ship would be in again. I just had to tear myself from her so as not to be too late in getting back to Lady Ross. The old lady said, "That girl has all faith and trust in you, Robert."

Wednesday, April 11, 1883

After breakfast at seven, Lady Ross threw her arms around me and kissed me, saying, "Good-bye, my boy, and God bless you." I got on board before eight o'clock and changed my clothing all ready for work in my new position in the same watch. Sam Hazzard, our third mate, and Mr. Roderick, second mate, got on board at nine o'clock. The only drunk in the whole crew was Grinnell.

Called all hands and got the running rigging down on deck from the pins and tops, the gaskets all off and the sails all loose. Manned the windlass and sheeted home both topsails, hoisting them with a chantey, "Way Rio." When the captain came on board at ten o'clock, we got the anchor to the cathead, set sail and with lookouts at fore and main mastheads, were off once more to the south to hunt the mighty sperm whale. Mr.

AFRICA, AND BACK TO ST. HELENA

Gifford, the first mate, and Mr. Roderick, the second mate, have all night in, while Mr. Hazzard, the third mate, and I, the fourth mate, head the watches at night. Sam Hazzard heads the port and I the starboard watch. I take my first watch tonight from seven until eleven, and Sam from eleven until three, then I take the deck until seven or breakfast. Sam eats at first table for breakfast, but at dinner I eat at first table. The cooper and steward always eat at the second table. Sam and I have the same stateroom and will get along fine together.

Of course Sam will be on deck while I am below and I will be on deck while Sam is below, except when all hands are called, or at dog watch when almost all hands are on deck, talking, playing, or singing.

Sunday, April 15, 1883

This morning we ran down toward St. Helena, hove to and put the captain on shore. He and I went up to see Mr. Thatcher, the American consul. The consul had both the cabin boy and Frank Knowles who had run away. The captain told me to take them both on board and not to punish them, and to tell Mr. Gifford not to put them in irons. While waiting at Mr. Thatcher's, I met his daughter and was talking to her when Captain Howland came into the room. He said to her, "Do you want to steal my right-hand man?" Mr. Thatcher laughed and said, "She would in a minute." I replied, "There is no fear," and they all laughed.

I took both runaways to the ship and returned to shore to wait for the captain, who had shipped a new St. Helena man.

Frank Knowles seemed glad to get back to the ship. I asked him why he ran away. He said that he did not know, just sort of got the notion. The mate gave the two a good lecture and told the cook to give them something to eat.

It seems that when the ship left the island the other day, it was to let the runaways think it was off for good, so that they would come out of their hiding places. It worked all right. Men who run away throw themselves on the American consul's hands, who is supposed to feed them and send them home on some vessel. The men can then claim that their ship sailed without them, and the consul has to look after them.

Friday, April 20, 1883

Today Sam Hazzard asked the mate if he could have the starboard boat put on the bow boat cranes because he liked a bigger boat and because I wanted to have my old boat back again. Mine is a little shorter and narrower but is mighty handy. We made the change when we hauled aback at sunset and hoisted them in their new places. My crew was much pleased at the change.

Today I had Fernand look to the irons and lances to see that they were clean and whetted sharp and in good order; to have the knives and hatchets sharp, and to have all the gear neat and in the right place, because it is the captain's boat and the one that he is most likely to use when going gamming. I did not want Captain Howland to find any fault.

Tuesday, May 1, 1883

For a week or more we have sighted nothing but ships, no whales. Today, the New York packet ship *Virginian* spoke to the captain through a trumpet. It was my first watch on deck tonight, and as we are right in the ship track I had to keep a sharp lookout. About ten o'clock I walked forward and told the man on watch to look sharp and sing out as soon as he saw a light. About midnight, standing at the mizzen rigging, and looking ahead, I saw a big ship coming right for us and no lights on her. I called to the watch to jump lively and brace the main yard around and told the man at the wheel to raise up the wheel lively hard over. We got out of the way all right but the captain came on deck and wanted to know what was up. I pointed to the big fellow going past our stern with no lights burning, and told him that I had seen her just in time. He said, "Well, Bob, you did not lose your head anyway."

Wednesday, May 2, 1883

At noon, took a sight with my old quadrant, and worked it

out as Lat. 17° 42' S. Long. 02° 02' W. Mr. Gifford asked, "What did you make it, Bob?" I told him and he said it was good, only a mile different from his figures and he has a good sextant. He asked where I got the old quadrant. I told him that I had paid ten dollars for it in New Bedford. He wondered why I had not bought a good one, and I replied that it cost all the money I had at the time and I wanted to practise while up in Greenland. The quadrant works all right only I have to adjust it every time I use it. Mr. Gifford said, "It is the one Noah had in the ark."

Sunday, May 6, 1883

While all the officers and harpooners were sitting around the main hatch and the vise bench, telling about the good times they had in St. Helena, Mr. Gifford asked, "Bob, why didn't you get in with that daughter of Mr. Thatcher's?" I said that I did not care for her at all. The mate went on, "She will have quite a pile of money some day as her old man has quite a place and is very well off. I said, "I don't want any grog shop money. I would rather have an empty pocket. As for the girl, I fear she would shame me." The mate said, "Well, Bob, she certainly hung around you the day she was on board." I said, "That may be so, Mr. Gifford, but you threw her on me to get rid of her and she certainly was one clip. The old man is fine but the girl is spoiled and has her own way all the time."

We talked and joked with one another all afternoon. I said, "Grinnell might fit in, as he is a good customer for the bar, but my trade is nil."

Monday, May 7, 1883

This day came in with fine weather and light trades. At daylight we set sail and braced up on the wind. Broke out slop chest for the men forward for clothing, stockings, and shoes. I made some pants and a jumper for the cabin boy, a good boy but thin and a half caste. I made the jumper with a large collar trimmed with blue and two blue bands on the cuffs and a star on each corner of the collar. I called him to try them on and they fit fine. How proud and thankful he was, for he had never had any clothes like that before! I told him to take care of them and to keep them clean. The captain was pleased and gave me a thin blue suit to cut down and make him another when I got time. The boy does many little jobs for me that he does not need to do, such as making my bunk, hanging the bed-clothes up in the boat to air, and he is right smart and very mannerly. He has no father and his mother is rough and drinks.

Wednesday, May 9, 1883

Sam Hazzard and I have both been working on the cabin boy's suit of clothes. Sailor-like, we make double seams, the same as we are used to when sewing sails. Frank helped us by threading the needles and waxing the thread. The cooper asked if we were making a tailor shop out of the vise bench. "No," said Sam, "It's a ladies' cooper shop for building clothes."

Sunday, May 13, 1883

Mr. Gifford is quite poorly today so the captain asked me to take the sun at noon. I worked out the latitude as 14° 55′ S. At three o'clock he gave me Mr. Gifford's sextant and asked me to take a sight for him. "All ready," he said. I sung out when I got the sun on the horizon. I gave him another sight, the captain giving me the time. He said, "Work it out and see how near you come." We both came out alike, Long. 00° 27′ W. The captain said, "You are very careful about the declination and that is what counts."

Fernand was telling me what a good time he had at St. Helena with the money I gave him from the boat race. He told me that he bought a suit of pilot clothes for eight shillings, dirt cheap. He bought it from one of the sailors from the gunboat's crew. He showed the suit to me. It was first class and would cost thirty-five dollars in America.

I had to laugh today when one of the monkeys stole one of the men's hats. When the man tried to get it, the monkey ran aloft and out on the end of the yard, the fellow after it. Then the monkey slid down to the deck on a rope.

AFRICA, AND BACK TO ST. HELENA

Sunday, May 27, 1883

Yesterday we ran into a school of blackfish that were playing all around us. Lowered four boats and gave chase. After three hours we got disgusted and gave it up, for we could not get near enough to dart them.

This morning we sighted a school of sperm whales and all four boats took after them. Mr. Gifford got fast and had a time killing it. His whale raised the Old Harry. It finally stove the mate's boat pretty badly. We pulled alongside and towed his boat back to the ship where we hoisted it on the davits and pulled the plug out. Mr. Hazzard got one small whale and Mr. Roderick another. These whales gave us ninety-three barrels of oil all told.

Sunday, June 3, 1883

The captain said, "Bob, tell the steward how to roast the fresh ham like you did at Fayal." I told the steward to score it and cross-score it; then to stick cloves in it, covering all with a thick paste; then to keep it well basted as it cooked and cook it plenty until the paste was quite brown; and to put some large sweet potatoes in the pan but not burn them.

The steward did a good job and most of the boys ate too much. Captain Howland talked to me for half an hour and asked if I ever drank anything in the liquor line at any time. I told him that I always dreaded it and kept from it, and as for tobacco, I said I never had a hankering after it. The captain said, "Good for you, Bob, stick to it."

Tuesday, June 12, 1883

Every day for two weeks we have been chasing humpbacks but they were wild and would not let us get near to them. Yesterday Mr. Roderick got his irons into one but they pulled out before he could kill it. How that whale did go, just as fast as an express train! Today Sam Hazzard got fast and got a lance home good and deep. That fixed that whale in good shape. We got the whale all cut in by two o'clock in the morning. We had to keep going to save it from the sharks, which were around the carcass by the hundreds. In fact, we killed over forty sharks while cutting in the whale.

Sunday, June 24, 1883

In the last ten days we have lost several whales, only getting two cut in. We see whales every day and give chase but they won't let us get close to them. Today we did not lower any boats and the men are washing their clothing and getting things fixed up. I held school for an hour for Frank Gomez and John Brown as I try to do every Sunday.

Sam Hazzard was trying to figure what he was going to get out of this voyage. He asked me how much money I had and if I spent the money from the boat race. I told him that I had not spent my winnings, that I had never drawn any from the captain and that I had eight dollars, plus twenty-five dollars for bounties, twenty from the race, and thirty dollars for painting the ships' names.

Sunday, July 8, 1883

In the past two weeks we have killed and cut in three more humpbacks. There are lots of whales. Hardly a day passes but we see them. To catch up to them is a different matter. The captain went gamming today on board the ship *Milton*, an old packet ship of over four hundred tons. Her windlass is abaft the foremast, the only one I ever saw that way. She is very roomy and square built. Her captain, Mr. Potter, is our captain's brother-in-law, and his wife is on board. Her mate is a bluff old fellow named Conley from Vermont. They swing five boats on her sides.

Friday, July 13, 1883

This morning, while cutting in a whale, there were lots of natives around the ship in their dugouts. The captain and mate cut big chunks of meat from the carcass for them. When their canoes had all they could carry, away they went singing and laughing, to the sandy beach where we could see them start a fire under a large caldron. The captain, looking at them through his glasses, said that they looked like the witches in *Macbeth*, dancing around the fire.

Saturday, July 14, 1883

This day was clear but for a heavy dew. The sun came out hot. I had to stay on board boiling oil, with my crew cutting the blubber up into horse pieces and firing the pots with scraps. Fernand, my harpooner, stood alongside of me at the hot pots and made the ship look like a steamer with smoke belching out of both smokestacks. The oil came fast. The heat in front of the fires was some hot in the middle of the day when the wind died out, and the smoke kept coming in our faces. When the sea breeze sprang up in the afternoon, it was a big relief and felt good.

MAYUMBA

Saturday, July 21, 1883

The ship anchored off Mayumba this morning. The boats pulled ten miles to the south but sighted nothing. We watched the natives fishing from their dugouts about two miles out from the beach. One fellow had his dugout as full of fish as it would hold. None of the natives had any clothes on except a grass apron. These blacks are large and strong, broad shouldered and deep chested, and not bad to look at. They are good natured, usually happy and quite peaceable.

Saturday, August 4, 1883

Chasing whales in the hot sun every day. Sometimes it seems as if they were just playing tag with us. They will let us get just about so close and then away they go. In the past two weeks we only got three whales.

Today my boat got alongside of one and Fernand drove both irons into it. I had put the boat right on its back. The whale smacked the water with its tail, soaking all hands and filling the boat. I slacked line and kept away until Fernand and I changed ends. I told Fernand to lay me on to the whale well forward. The whale was lying still as if dumbfounded. I got a lance into it three or four times and set it spouting blood. I yelled, "Stern all, stern all, and slack line." We backed away about fifty yards. What a time it had plunging and breaching! It ran back and forth like a race horse but at last rolled over on its side, fin up.

Friday, August 10, 1883

While we were stowing oil down in the lower hold, the captain saw a whale coming and said, "Bob, get that fellow. There's five dollars for you if you get it." The captain and cooper helped lower my boat, the other three boats being out already chasing whales. With sail up and all my crew on the weather gunwale paddling for all they were worth, we were fast to the whale in about twenty minutes. It went like a race horse to the windward and shook the boat almost to pieces. Then it headed toward the ship as hard as it could go. When it slackened speed about a quarter of a mile from the ship, all hands hauled line. I got a lance into it and it started to have a circus all to itself. It rushed here and there, keeping me and my crew busy trying to stay out of its way on account of the reach of its ponderous tail and fins sweeping the water all around. It raced and breached and rolled over and kept on rolling over and over quite near the ship all the time. It spouted blood but I had to keep back until it wore itself out, or it would have smashed the boat to splinters. The men rested on their oars ready to pull or stern at a minute's notice. At last my chance came. The boat was on the whale in no time. My harpooner laid me on near the fin. I lanced it good and plenty abaft the shoulder blade. Blood came bubbling out of the spout holes like liver. I sung out, "Stern hard and slack line." The whale went into a flurry. We were none too soon. Its tail was lashing the water to foam. It rolled over and over, gave a mighty lunge forward and died only two boat lengths from the ship.

At supper time the captain was telling Mr. Gifford how my whale wanted to come on board the ship. He said, "Bob met it and you should have seen that boat of his go with the crew on the weather gunwale paddling for dear life."

Tuesday, August 14, 1883

Yesterday the waist boat got fast to a whale but could not get near enough to kill it until nearly dark. Then the whale raised Cain and stove two boats. The worst of it was, when we did kill the whale it sank. We tried to raise it but could not, so had to lay all night at anchor where the whale sank, until morning when the ship was brought to the whale. Even that did not do any good, so the sharks got it, and what a feast they had!

Today Sam Hazzard got fast to a whale late in the afternoon. All four boats fastened to it so as to finish it quickly, but they were dragged around by the whale as if they were nothing at all. When it stopped running, all four boats got around it and quickly killed it.

FROM ST. HELENA AS MASTER OF A SHIP TO BOSTON

Chronometer wrong—Consul requires navigator to take Daylight *to Boston—Ferguson chosen master—Signing ship's papers—Shorthanded crew—Officers dead—Storms—Black eye—Irish bully knocked out—Yarn of mutiny on clipper—Arrived at Boston—Owner of barque—Guest at owner's house—Friend at Custom House—Ferguson offered captaincy of large ship—Decides to return to Kathleen—Driving with owner's daughter—Services at Gaelic church.*

CHAPTER X

FROM ST. HELENA AS MASTER OF A SHIP TO BOSTON

Thursday, August 23, 1883
The day before yesterday we got two sperm whales that gave us forty-eight barrels of oil. This morning we ran into the harbor at St. Helena and let go the anchor. The mate told me that the reason we came here this time was because the chronometer had gone wrong. The captain went on shore to try to get another before going to the African coast again. He came off with one that came from a condemned ship that was broken up here.

Captain Howland and the mate talked for a while and then both went on shore. Mr. Gifford came on board before supper, called me down in the cabin and said, "Bob, do you see that barque lying there? How would you like to go captain of her to Boston? There's a chance if you will take it. I think you will get the chance as there is no man on the island to take her home. The American consul has her on his hands and has asked Captain Howland about letting him have a navigator. The captain said that you could do it. It seems that both the captain and the mate of this barque died at sea, and her third mate came in here and died three weeks ago. It will be two months before the consul can get anybody sent out from Boston. I told him that you were the only navigator that we could spare. He asked me to speak to you. What do you say?"

I asked, "Has the crew all gone or are they still on board? Before I give you an answer, Mr. Gifford, I should like to look at her and see what kind of a craft and crew."

Mr. Gifford and I then went on board the barque and looked her over. She was a good-looking craft with double topsails, named the *Daylight*, and looked as if she would be handy. Her gear was good. The ship was sound, solid, and not very old. From the steward, who had been on her for several voyages, I found out that she was a good seaboat. The crew was mixed, and none of them knew anything of navigation. This barque was 360 tons registered and a good mold. I told Mr. Gifford that I would make out and that I would do my best.

After going back to the *Kathleen* for supper, the mate had me put him on shore and wait until he had told the captain that I would go on the barque. After waiting an hour, the mate came and we went back to the *Kathleen*. He told me that I was to go to the consul for the ship's papers, to get all instructions and a letter to take to the owners. He told me to get everything ready to get under way tomorrow forenoon. Mr. Gifford said, "Bob, do not take baggage with you, only a change of underclothing. If you need any more clothes, get them in Boston. From there, go to Liverpool and take the mail boat that stops here. If all goes well, by the time we come here for letters in October, you should be here."

When I got to our little stateroom, Sam was worried. He thought I had got into trouble. I told him what had happened and he said he could sleep better now. I said a prayer, turned in, and after quite a while, slept.

Friday, August 24, 1883

I could not express my feelings when I came on deck this morning. The first thing I saw was *that* barque. After breakfast, Mr. Gifford called the boat's crew to put us on shore. We went straight up to Mr. Thatcher's. Captain Howland was there. He gave me instructions and some good advice, cautioning me about my navigation; about the care of the chronometers; to watch closely for sudden squalls when near the Line; to keep my eye on the barometer and to have all things handy in case of a blow.

Mr. Thatcher got me to put my name on the ship's papers as master, gave me some letters to the owners at Boston, and told me how to explain matters to them. He told me to come to his house on my return to wait for the *Kathleen*.

Captain Howland shook hands with me and told Mr. Gifford to give me a hand to get under way. The mate and I went back on board the *Kathleen*. I got my little bundle of things. I did not take my old quadrant for there were two good sextants on board the *Daylight*. Mr. Gifford took ten of our men over to help get under way and hoist the sails and anchor. The American consul was already on board. He called the crew into the waist and told them that I was their captain and for them to obey all commands willingly and to stand by me. He then shook hands with me and wished 'me good luck. Mr. Gifford saw that the anchor was inboard and lashed. We got the sail all on including the topsails, topgallant sails, royals, and all the light sails. Mr. Gifford, being in no hurry, said that I had a good crew but to look out for a certain reddish-looking Irishman who was inclined to be a bully. The mate said good-bye and all the Portuguese sang out, "Good-bye, Captain Bob, we wish you a good passage." They had about four miles to pull back to the *Kathleen* but that was nothing for those lads.

I gave the course to the man at the wheel, north-northwest, and we were soon going through the water at a good pace. As the men coiled up the rigging, I watched their movements and the way they did things. They did not know that I was taking stock of them.

When the deck was all clear, and the cable down in the chain locker, I called all the men aft and got their names. There were ten all told. I picked one, Anderson, the Norwegian, to act as mate and told him to move his things into the cabin. The men were divided, four men to a watch, one mine and the other the mate's. Tonight I took first watch. There were two Swedes, one German, one Norwegian, one Austrian, one Englishman, one Irishman, and two Americans in the crew, not counting the steward who was also an American.

I told my mate that I did not intend to shorten sail at night as I wanted to make quick passage to Boston. This Anderson is a first-class sailor, but he has never had charge of a deck before. I told him that he knew what to do, but not to let the men get too familiar and to keep them in their places. He told me that Ryan was a regular bully and had been a trouble maker all during the voyage. I told Anderson, "He is in your watch. If he does not obey orders, let me know and I will attend to him." Anderson said that the rest of the men were all able seamen, none of them shirkers. At eight bells I set the watch, one man on the lookout forward and one man to relieve the wheel.

I was thinking what a change this was from last night, when I was on a whaleship of 314 tons with a crew of thirty-five men all told, and tonight captain of a merchant barque of 360 tons with eleven men all told. However, I have been shorter handed than this and pulled through. When the mate came on deck I gave him his orders and went below to sleep until four o'clock or to think over this new situation.

Saturday, August 25, 1883

The mate called me at four o'clock and reported all well. A man came to the wheel with a big cud of tobacco in his jaw. I told him that there would be no tobacco at the wheel as there was time enough for that when forward. I went forward to see that the lookout was wide awake and if all the lights were burning bright. At eight o'clock I had a talk with the steward about what food there was on board and how he fed the men. I told him to give them plenty and as good as he could, but not to waste it. If he ran short, I said that we would stop at Barbados and get more. The steward is a first-class cook and has a splendid, well fitted-out galley.

I figured our position at noon as Lat. 10° 58′ S., Long. 12° 36′ W.

At dogwatch we passed a ship and a barque racing with each other. They carried stunsails but were deep in the water. The barque was the *John Wright* of Bangor, Maine, bound for Baltimore. Anderson said he had made a voyage to China on that barque three years ago.

ASCENSION ISLAND

Sunday, August 26, 1883

I sent a man aloft to see if he could see Ascension Island. It was four points on the starboard bow. At noon we were abreast of it in Lat. 7° 58′ S., Long. 14° 30′ W. This island is low lying but almost all rock. The British Government uses it for a coaling and signal station. There is no water on the island but that made from a condenser. A gunboat is stationed here all the time.

One of the Castle Line steamers passed us today, and we sighted a large school of sperm whales. Anderson and I had a long talk. I found out that he had been at sea ever since he was fifteen years old, mostly on British and Yankee ships. He had a fair education so I advised him to take up navigation and then he could advance himself. I told him to watch the barometer, to look out for squalls, as we were getting near the Line, and to call me if he had any doubts about the weather.

Tuesday, August 28, 1883

The weather was hazy this morning and I did not like the looks of things, so I napped in a chair on deck until noon when I took an observation of the sun and another at three o'clock. Made our position Lat. 02° 50′ S., Long. 22° 58′ W. There was a light squall in the afternoon with wind and rain. Took in the fore and main royals and fore and main topgallant sails. Furled the royals but set the topgallant sails again after the squall.

Wednesday, August 29, 1883

Wind very light. Running close to the wind with the yards braced up sharp. Saw three ships close together, becalmed, about five miles off. At five o'clock we had such a heavy downpour of rain, but no wind, that the scuppers could not let the water off the deck. The deck was flooded nearly a foot deep with water sloshing around as the ship rolled on the swell. I stayed on deck all night.

Thursday, August 30, 1883

Today is scorching hot but there is a good breeze and we are going right along, hoping to get through the doldrums. Made out we were Lat. 0° 43′ S., Long. 26° 30′ W. If the breeze holds out, we will be across the Line by supper time. I was napping in a chair when a clap of thunder woke me about three o'clock. Called all hands on deck to shorten sail. Clewed up topgallant sail, hauled down gaff topsail, flying jib, clewed up mainsail and foresail, and let go fore and main topsail halyards. Furled the light sails and brailed in the spanker.

When the storm broke we were ready for it. The men proved themselves all able seamen, doing what was needed, with the mate working along with them. We kept her off before the wind and heading her course. She heeled over and dove into it until she buried her lee cathead, so we dropped the foresail. I took the wheel myself and told the man who was already at the wheel to help. The roar of thunder was deafening, just one awful roar. The lightning flashes were blinding and almost continuous. The rigging screeched and shrieked from the high wind. It was dark as pitch between the lightning flashes. Set two men on lookout watching sharp ahead when the lightning flashed and told them to bawl out if they saw anything. The squall lasted until ten o'clock. At midnight we set fore and main topsails, mainsail, spanker, and jib. An hour later the stars came out except the Southern Cross, which was now below the horizon. We may look for the North Star any time now.

Saturday, September 1, 1883

Yesterday we boomed along at a fine pace. This morning, there was a barque a short ways ahead of us so I had the fore and main royals set. Then we gained on the barque until I could read the name *Bengal* of Bristol, England, on the stern. We came close to her and spoke to her captain, who said they were bound for New York. I got a sight at noon and made out Lat. 3° 35′ N., Long. 31° 06′ W. Tried the pumps and found very little water.

Sunday, September 2, 1883

All sail set, heading on our course and booming right along, about ten knots by the log. Here I am pacing back and forth on the quarter deck and feeling proud, but anxious to get north into cooler weather, for the sky is clear and the sun very hot. I was just thinking about all these men being much older than I am. Worked out the position today as Lat. 5° 52′ N., Long. 35° 32′ W. I am hoping this wind will last and give me a chance to reach the trade winds.

At eight bells one of the men came to relieve the man at the wheel. He had a black eye that looked bad. I asked him what

was the matter and he said he had a fight in the watch below.
I told him that I did not allow any fighting and that if there
was any fighting to be done on board, I would do it. He said
he did not want to fight but Ryan was always picking on him.
After supper, I called Ryan aft and told him that if he raised
his hands to any man on the ship, I would put him in irons for
the rest of the voyage and put him on bread and water. I told
him that he was a trouble maker, to go forward and not let me
hear any more of his tyranny over his shipmates. He went forward but held his head high and said nothing. The steward said
the Irishman was a bully and had fought with all the men except
the Norwegian and that he also used to slur the second mate.
I asked the steward if he could fight. He replied that the Englishman nearly whipped him one day and now he does not
bother the Englishman any more.

Monday, September 3, 1883
   We got into the trades today and are going right along in
Lat. 7° 57′ N., Long. 38° 06′ W. At eight bells Ryan, with a
pipe in his mouth, came to relieve the man at the wheel. I told
him to take the pipe out of his mouth and go leave it forward.
He went but did not come back. I knew then that the sooner
I took the Irishman in hand, the better. I told Mr. Anderson
to stand by. I went forward and called all hands on deck. When
Ryan came out of the forecastle, I grabbed him by the neck and
said, "You have run your rope." He laughed and tried to kick
me as I forced him amidships. I said, "Defend yourself if you
know how." He came at me like a wild bull, head down. I
stepped lightly to one side and flattened his nose all over his
face, making the claret flow. The blow staggered him. I saw
right away that he had no science. I feinted and played with
him. He thought he had an opening and sprang for me. He took
the bait just as I had expected. As he rushed, I sidestepped and
gave him one blow on the jaw just as hard as I could. His coming at me gave the blow double force. He got an awful crack
and crumpled up. He rolled clean over into the lee scuppers,
lying there like a log. We soused him with a couple buckets of

water but he did not come to his senses for fifteen minutes. When he did come to, I told him a few things. I said, "If it had been any other captain, a belaying pin would have been taken and your skull cracked; then you would have been put in irons for mutiny. I don't want to hear any more complaints about your being a bully. Try to act the man, not the brute. Go take the wheel and don't provoke my anger. By rights, I should have you put in irons down in the hold. If you will act decent and like a man, everything will be all right."

Tuesday, September 4, 1883

This day came in with strong trades, so strong that I was afraid something might give way so I took in the royals. The mate said that he was glad to see me knock the Irish bully out so easily, and that the crew were so pleased that they will do anything for me. He heard the Englishman say that the captain was clever and knew how to handle his dukes and that he would not want to run afoul of him. When Ryan came to the wheel at night, he asked if he could speak to me. I said, "Speak away." He asked if I had put it in the log against him. I said, "No. Attend to your steering."

Wednesday, September 5, 1883

All sail set and strong trade winds in Lat. 11° 40′ N., Long. 42° 13′ W. I told the mate to take a look over the rigging and to see that there was no chafing, in case we should get a hard blow in northern latitudes such as often occurs at this time of the year. The men touched up the rigging, chafing gear, and the battens on the swifters. We put new sheets and tacks on the foresail, and overhauled the reef tackles. The boat which set on top of the house got cleaned and painted. It's a big, heavy, clumsy boat. I told the mate to see to it that the mast, sails, and oars were all in it and that it had a keg of fresh water. There's nothing like having things in good shape. The steward says that the crew is in good spirits. He heard them singing today for the first time in months.

OF A SHIP TO BOSTON

Monday, September 10, 1883

The wind holds good and we are heeled over at a good angle leaving a straight wake astern. Talking with Anderson today, he said the worst ship he ever had been on was a Yankee clipper bound for China. Her officers were all brutes. They had a couple of mutinies. One night in a bad blow the mate was missing, having gone overboard. Another bad night, trying to get around the Horn in a snowstorm, the captain came near going overboard. One of the men knifed him, but the captain was such a big heavy man that he could not be thrown overboard. He came near dying. Some one clubbed the second mate while he was kicking at a sailor pulling on the lee main brace. It stunned him. After that it was a hell ship all across the Pacific. The men making up the crew were all old packet rats who stuck together and would not stand abuse. When near Shanghai, the crew set fire to the ship and took the big boat. Anderson had an awful time on shore for eight months and even had to steal to keep living. Finally he stowed away to another port where he shipped on a clipper to New York. Some of the mutinous crew were caught and got ten years' prison in Shanghai. One man who had been caught by two policemen was desperate and killed both police with one of their own guns. He was never caught again.

Thursday, September 20, 1883

The weather has been in our favor right along. Outside of speaking to a vessel or two there has been nothing exciting. The mate has gone over the rigging and everything is in good shape, in case we get bad weather. The men are all in good spirits. Ryan came to me at dog watch and wanted to know if I was going to tell the shipping commission how he had behaved. I told him to think no more of it and to try to be more agreeable to his shipmates, for they were all good fellows. Then I said that if he would make friends of all of the crew that I would not report him. He thanked me and was very meek. Ryan is a different man altogether. He has lost his dourness and is quite chipper. If I were to make a complaint to the shipping commission he would lose all or most of his pay, so I thought that he had punishment enough. For all of that, he is a good and able seaman.

Friday, September 21, 1883

By noon we neared the Gulf Stream, for there was seaweed floating in quantities and lots of jellyfish. The water too has become quite blue. The wind getting stronger, we took in the flying jib, gaff topsail, and topgallant sails. Had to call all hands to haul down and clew up and let the upper topsails down on the caps. We worked quickly but none too soon. We barely got the sail off her in time to save the spars. When the big wind came, it came with a bang and lasted about two hours. Anderson did the work of two men. His cheerful voice kept the men hard at work but willing. The weather was still fluky after the blow, and the glass none too steady. I kept on deck all night.

Saturday, September 22, 1883

The weather clearing, I told Anderson to watch as I was going below for a well-earned rest. We are in Lat. 35° 41′ N., Long. 69° 19′ W. and running free with all she can carry. Have lots of company in sails all around us, ships, schooners, large and small brigs, barques, and steamers. I counted eight schooners bound south, all in sight at one time. Told the watch to keep a sharp lookout and to keep our own lights bright.

Sunday, September 23, 1883

This is a perfect day with the air sharp and cool. I got a nap this afternoon. I fought sleep all night, for the shipping is getting so thick that I have to watch carefully. All of the crew are in fine shape, ready to jump at a moment's call. There is not a loafer in the lot. As the glass was falling, I told Anderson to get all the sleep he could as he would likely be called out on short notice.

Monday, September 24, 1883

At breakfast it was blowing fresh with variable winds, so we took in the light sails and furled them. Quite a heavy sea came on and we had to clew up the mainsail and foresail. I took the wheel myself and sent the man at the wheel to help with the heavy sails. Took in the jib, brailed the spanker, and got ready for a blow. All we had to do was let go the topsail halyards and all would be reefed down. At noon it was blowing a stiff gale and hauling around to the east. We are standing up under double topsails down on the caps. It was a hard squall and there was an awful sea on. The wind died out at night. The glass was rising. Called all hands to get more sail on, topsails, jib, foresail, mainsail, and spanker. That steadied her. At midnight I took a nap when I saw that the sea was going down.

Tuesday, September 25, 1883

Got all sail on at daylight with a fair slant of wind but in a drizzly rain. Sighted a lot of fishing schooners going in and out about Lat. 41° 38′ N., Long. 70° 15′ W. At two o'clock in the morning I sighted Cape Cod Light. Having been in Boston Harbor only once, I was afraid to risk going in at night. I shortened sail and kept her off until daylight. I stood on deck all night with Anderson.

BOSTON

Wednesday, September 26, 1883

Early this morning it was fine and cool with a fresh breeze. We headed in past Cape Cod and up the harbor with the flag at the fore, calling for a pilot. After clewing up the mainsail, we got a pilot. A tug came alongside and took hold of us. Furled the fore- and mainsail. The anchor was at the cathead, ready to let go if necessary. All light sails were clewed up. All hands were busy getting up hawser from the fore hold. The rigging was all coiled up off the deck while the tug was pushing us to the dock. The men worked like beavers. By five o'clock in the afternoon we got all fast alongside of the dock with the lines and spring line in place and the deck swept and businesslike.

A gentleman came on board and spoke to Mr. Anderson. He also spoke to the steward and shook hands with him while I was talking to the pilot. Crossing the deck, he came over and shook hands with me, saying that he was Mr. Aiken, the owner of the barque. He had been looking for her for two months and had received no report of her. Last night a steamer reported the barque, so he sent a tugboat to meet her. He asked how I came to have charge. I told him that the captain, mate, and second mate were all dead before I had come on board. He took it very hard and was shocked. We went down to the cabin and got the ship's papers. I handed him the letter from Mr. Thatcher, the American consul at St. Helena. He read it and turning to me, laid his hand on my shoulder and said, "Captain Ferguson, I am glad to meet you. Just lock things up and come with me or we will be late for dinner." I told him that I would have to shave and clean up but he would not have it, saying that he wanted to talk to me. I told the mate to look out for the ship, keep his eye on the men and not let them get drunk. I cautioned him also not to let any boarding-house runners on board.

When we got up to the dock, there was a horse and carriage waiting. Mr. Aiken and I jumped in and drove to the outskirts of the town to a fine large house. Taking me by the arm, he led me into the house and introduced me to his wife and daughter Priscilla. He showed me to the bathroom and told me to help myself to the razors. After a bath and shave, I hardly knew myself. Going downstairs to the dining room, Mrs. Aiken made me feel right at home. Mr. Aiken said grace. It was a fine dinner after the rough ship's fare. Mrs. Aiken said that I was a young man to be captain of a ship, as they generally were middle-aged men with gray beards. I told her that I was not a captain, that I only brought the *Daylight* home from St. Helena. She wanted to know all about the island. I told her about the people, about Longwood, the house in which Napoleon lived while he was prisoner there, and about the clear cold spring that gushed out of the rocks near his tomb.

Mrs. Aiken offered me a glass of wine, but when I did not touch it I had to explain that I never touched wine or liquor of any kind. Mr. Aiken said, "Well, let's go into my study and have a cigar." I laughed and told him that I could not smoke and never had tried.

In the study, I told him all I knew and how I came to bring the *Daylight* home to Boston. He thought that I had made a quick passage. I said that the *Daylight* was a good sailer and would lose no time if handled right.

He then asked, "Would you go out in her as master?" I told him that I could not as I had promised to go back to my old ship, the *Kathleen*, at St. Helena as soon as I could.

Just then his wife and daughter came in. Mrs. Aiken asked me where I belonged. I told her that my home was in Philadelphia. As she and I talked a great deal, Mr. Aiken had little chance. I was shown to my room and told that breakfast would be at nine o'clock.

Thursday, September 27, 1883

After breakfast Mr. Aiken and I went on board the barque and got the ship's papers to enter at the Custom House. The clerk at the Custom House thought I looked rather young for a captain. Mr. Aiken told the clerk that the captain whom he had known was dead, and that I had brought the ship home. The clerk kept looking at me. Finally he said, "Scotty, don't you know me? Don't you remember Dudley Davenport who was up in Greenland on the *Abbie Bradford*?" Then it came to me and we shook hands. He had been a grocer's clerk in New Bedford, and had gone north just that one trip, but it had been enough for him. He asked me where I had been and I told him. When through, we went to the shipping commissioner's place where the men were getting paid off.

On the way to Mr. Aiken's home for lunch, he took me into a store, bought me a new suit of blue clothes, a hat, and rigged me out completely. Afterwards, on the way down to the ship, I spoke to him about Mr. Anderson and said that he was a good man and worth retaining. Mr. Aiken later told Mr. Anderson that he was to ship as second mate and to stay on board and take care of the ship until he could get a captain.

When I bid Anderson good-bye, saying that I was sure he would get along all right, he thanked me and cried over me like a child, even though he was a great big fellow. The steward was going ashore to his family and was wishing that I could go and spend a few days at his home. At dinner tonight Mr. Aiken told his wife how I had met an old shipmate at the Custom House and how we both had wintered up in Greenland in '76.

Mr. Aiken still wants me to go out as captain of the *Daylight* or another still larger ship that he has, but I told him that it would not be fair to Captain Howland, since I had given him my promise to get back to my ship. I went up to bed at ten o'clock, but before turning in tried on my new clothes.

Friday, September 28, 1883

Shaved, dressed in my new clothing, and had breakfast. Mr. Aiken said, when I was leaving his home to go down to the ship, to be back at noon for lunch as he was going to find out about my passage on the Guyon Line to Liverpool. I went down on board and found Anderson busy getting ready to discharge cargo. By noon the stevedores were at work. .

At lunch Mr. Aiken said he would get my ticket for me when he went to the office. He said that Priscilla would keep me busy for the afternoon. His wife and daughter asked me many questions about the ship and what kind of a passage I had coming here. Priscilla said her father had told her that there were no officers on board and that I had been on deck for two days and nights without any rest. She also said, "Papa wishes you would go out in one of his ships, and why not?"

At dinner Mr. Aiken told me he got a ticket on the *Arizona* for Monday and a ticket on the Castle Line from Liverpool to either Capetown or St. Helena. The agent said that I would probably catch the *Dun Robin Castle* and lose very little time.

Mr. Aiken asked me to tell some of my experiences and adventures, so I told of things that had happened in the Arctic, how the natives lived; what a thrill it was to be dumped in the icy water by a whale upsetting your boat; how we hunted bear, seal, and walrus, how we hunted deer up in the hills; how the natives built their snow houses; what the natives ate and what their skin clothing looked like. I followed that by telling of our cruise off the coast of Africa, Madagascar, and the wonderful

Seychelles Islands. They would have sat up all night questioning me about places but Mr. Aiken said, "Bedtime."

Saturday, September 29, 1883

This morning I went down to the wharf on board the *Daylight* to talk to the second mate. He said most of the men had come on board and wanted to know if I was going out as captain again, because they wanted a chance to go out with me. He said, "Even Ryan came down and said he wanted to go out with you." On my way back, I got lost and had to get a policeman to direct me. The streets seemed to run around and around, but I finally got back to Mr. Aiken's for lunch.

Mrs. Aiken said that her daughter was going for a drive that afternoon and asked if I would like to accompany her and see the country round about. I said I would and thanked her. We went in a light buggy with a fine horse on a beautiful drive toward Lynn. There was no dull time as this girl was a fine talker. She asked me about Philadelphia and its suburbs. I told her about Germantown, Chestnut Hill, and Bryn Mawr. She said that she hoped I would stop for a week and that she enjoyed my company.

At supper Priscilla told her father what a pleasant drive we had. In the evening all three wanted to know more of the places I had seen. It was midnight before we knew it. Mr. Aiken said a prayer and was glad when I told him that I liked to go to church.

Sunday, September 30, 1883

After bathing, shaving, and dressing, I came downstairs at nine o'clock on the tap of time and had a splendid breakfast. When through, Mr. Aiken had a smoke in his study and said, "It's a queer thing that you don't smoke." I said, "I come of a large family of eight boys none of whom smoke, but as for my father, he is a heavy smoker, generally using a black clay pipe." He asked what denomination I was. I said I was a Presbyterian or blue stocking as folks call us down here, so we went to a Presbyterian church, the first I had been to in four years. I told

Mr. Aiken that there were none to go to where I had been except Catholic, Mohammedan, or Episcopal and that they were all good and steered the same course, bound to the same port, only a different chart. We heard a good sermon that held our attention and went home to lunch. I asked Mr. Aiken if there was a Gaelic church here. He said there was, but wondered if I could understand the tongue. I told him that it was my native tongue and that I spoke it altogether when I was a boy. He called his wife and Priscilla and asked them if they knew where the Gaelic church was. Priscilla said it was only about half a mile away and suggested that as there was evening service that we walk over after supper. On our way she asked about the natives in Africa and how they dressed. I said, "They don't require much dress and never overdress at any time but often underdress with only a few strips of grass." "And how do the Eskimo women dress?" she asked. I replied that they dressed all in furs, mostly seal, fox, and deerskin.

When we came to the church, a Highlander ushered us in and spoke in Gaelic. It was a large church with a big congregation. The sermon was just the kind I heard when a boy, all fire and brimstone. The Lord's prayer was said in Gaelic, with everybody standing. When we came out and heard groups of people talking the old tongue, I thought for the moment I was back in the glens of the Highlands. It reminded me of days in Scotland long gone by, when Dr. MacCloud thundered out a sermon that made me tremble with fear and dread. Priscilla noticed that I had felt the sermon. I told her that if she had understood it, her attention would have been held from beginning to end. When we got home, Mr. Aiken asked her if she enjoyed the sermon or the walk most and she said, "The walk, of course, because I did not understand one word of the sermon."

FROM BOSTON TO LIVERPOOL, ST. HELENA AND ST. PAUL DE LOANDA, AFRICA

On steamer Arizona—River Mersey—Liverpool docks—On Steamer Dun Robin Castle—Dancing and music—At St. Helena—Guest of American Consul—Prefers to stay with Lady Ross—Donkey ride to Sandy Bay—Three beach combers—Ferguson obtains a yarn from a beach comber—The Kathleen arrives—Welcomed by shipmates—A talk with the Captain—Leaving St. Helena—Four whales from one school—Nearing St. Paul de Loanda.

CHAPTER XI

FROM BOSTON TO LIVERPOOL, ST. HELENA AND ST. PAUL DE LOANDA, AFRICA

Monday, October 1, 1883
After breakfast I went into the study with Mr. Aiken, who gave me my passage tickets and three hundred dollars for wages. I thanked him and bade good-bye to his wife and daughter. I told them that although I had been here but a short time, they had made me feel very much at home or like an old friend and I thanked them. They said it had been a pleasure to have had me with them.

Mr. Aiken took me to the tender for the steamer. We shook hands and again he wished that I was going out in one of his ships. The steamer lay two miles out. After I got in my stateroom, I sat down to think and wonder if it would have paid to have accepted Mr. Aiken's offer and lose all of my share of the catch in the *Kathleen*.

When we passed out beyond Cape Cod, it was blowing a gale with quite a nasty sea on. I noticed quite a fleet of fishing schooners all under storm trysails and riding the gale like a lot of ducks in a millpond. In the evening I went to my stateroom and thought of Sam in our little cramped stateroom on board the whaling barque *Kathleen*, and wrote up my diary for two days back.

Tuesday, October 2, 1883
Out on deck at six o'clock for a look around, fore and aft. The *Arizona* is a fine large steamer with every new device and contrivance to make work easy for man. We passed one of the Inman Line with a good figurehead, clipper bow, a red funnel, and a row of boats along her upper deck. There do not seem to be many passengers on our boat or else they are seasick. After dinner I went into the social hall and read a book until I became drowsy, then took a stroll on deck but there was not a soul out there.

Friday, October 5, 1883

Until yesterday, I had not spoken to anybody on board except the steward, when a man who had been over in Canada came over and we had a good talk together. He was going back to England and had been seasick ever since he came on board. Today the fourth officer talked to me. He had been mate of a sailing vessel and said that even a captain of a sailing vessel could only get a second or third mate's billet on a big steamer. I told him that I was going to St. Helena by the Castle Line and he said he would find out for me just when my boat sailed. Time goes slow with me and these few days have seemed like a month, but day after tomorrow we will be in Liverpool.

Saturday, October 6, 1883

I found out that my Castle Line boat will sail at two o'clock Monday, so I'll not have long to wait. There have been lots of ships in sight today, going and coming in all directions. Probably more ships go in and out of Liverpool to all parts of the world than at any other port. Orders from the bridge came to get all boats swung out and ready to lower at a moment's notice in case of an accident in the Channel.

LIVERPOOL

Sunday, October 7, 1883

As we went up the River Mersey today, the scenery looked familiar to me. Along the bank were landmarks that I had looked at many times when I was quartermaster on the *Pennsylvania* seven years ago. We got in to dock in the Canadian slip at three o'clock, high tide. Most of the passengers went on shore but I stayed on board because I could not get on the other steamer until tomorrow. Later I went to the fourth officer to ask where the Castle Line docks were. He told me they were up north by the African docks and gave me a pass to come on board again. There were two Castle Line boats there, the *Balmoral Castle* and the *Dun Robin Castle*, but I did not go on board either of them. However, I noticed a sign saying that the latter, my boat, would sail at noon sharp, on Monday.

ST. HELENA AND ST. PAUL DE LOANDA, AFRICA

Monday, October 8, 1883

As I shook hands with the fourth officer, he wished me good luck and asked me where my trunk was. I held up my bag and said that was all I had, as my sea chest was on board my ship and that I traveled light as I did not wish to be encumbered. On board the Castle Line steamer, a deck steward showed me to my stateroom, a fine airy room well fitted, all gold and white, and told me that I could have lunch in the dining room. The tables in the dining room were well filled with passengers.

Saturday, October 13, 1883

Outside of a rough trip going across the Bay of Biscay which made most of the passengers sick, it has been a quiet trip. At night there is music of all kinds on deck and lots of singing down in the saloon. Some men and even the women play cards and gamble. I have been all over the boat, including the engine room. These boats are fine, with every convenience for comfort, and make their own ice on board.

Today came in with a heavy thunderstorm, the thunder rolling along with a heavy deafening sound and the lightning sharp and wicked. Thinking that we must be in the neighborhood of the Cape Verde Islands, I spoke to one of the officers, who said that we had passed a little to the westward of them about noon. He asked me where I was going and I told him to St. Helena. He said he had been there about thirty times but had never gone on shore because they only stopped long enough to deliver mail.

I keep by myself among all the passengers and have many a laugh watching the moves of some of them. I like to watch the dancing, sometimes a sailor's hornpipe or an Irish jig. I enjoy listening to the music and the songs. Tonight there was a big Irish girl who was anxious for a dance. She was good. We danced four sets together, then I went up on deck to cool off.

Sunday, October 14, 1883

This morning the sea was as smooth as glass, and the steamer hummed right along. We had church service in the saloon for an hour. The purser gave a good talk. Two of the quartermasters sang hymns with the passengers helping in the singing.

Monday, October 15, 1883

Very hot weather today with light winds from aft. I looked around for a shady nook forward where I could see and yet get a breath of air. All of the passengers were on deck in chairs, those who could get them. Some had fans and were busy keeping them going. I feel the heat here more than on the sailing ship. At six o'clock we had a heavy downpour of rain, as heavy as I have ever seen, but it cleared up in an hour and the moon came out, bringing quite a few of the people on deck. There was a concert in the saloon where one little girl about fourteen years old, and one who I thought was her mother sang, "Rolling Home Across the Sea." I never heard anything like it. The little girl had such a strong voice that it made the saloon ring. The people clapped their hands and praised her but she sang no more this night.

Thursday, October 18, 1883

We struck the southeast trades today, which gave us a strong cool breeze. Ascension Island was passed and tomorrow will bring us to St. Helena. The second officer asked if I thought my ship would be there waiting for me. I said, "Hardly, but she will come in for letters sometime soon." I told him that we expected to go back to the west coast of Africa near St. Paul de Loanda, for that was the agreement among the fleet of whalers.

ST. HELENA

Friday, October 19, 1883

After lunch I packed my bag ready to go on shore, but it was half-past two before we sighted the island. We were in and anchored by four o'clock when I went on shore on the lighter with the mailbags. With my bag in hand, I walked up to the American consul's office. As I handed Mr. Aiken's letter to Mr. Thatcher, he looked at me and said, "You took the *Daylight* to Boston, but you look different." I answered, "Just a clean shave and a fine suit of clothes." The consul shook hands and told me to make myself at home until my ship came in. He said that the place seemed dead, not a ship in the harbor. He thought I had made a quick passage to Boston and wonderful time coming back. He called his daughter to show me to a room. She did not know me at first but later realized that I was the man who had showed her over the *Kathleen* about a year ago. She sat down on the bed and asked many questions. Bold, and quick as a flash, she grasped my hands and said, "Aren't you going to shake hands with me?" As she left, she said that supper would be ready in half an hour. Mr. Thatcher was very nice to me, but that daughter of his was a bold lassie and outspoken before her father. She certainly was spoiled. After a splendid supper Mr. Thatcher went out to smoke and asked me to join him, but I told him "No." The girl called me inside to talk to her. She sat as close to me as she could and said, "Bob, that is what Mr. Gifford called you, you are a good-looking chappie. Why don't you stay on the island and live on shore?" I told her that even if there was work for me, I would soon tire of it as it was a rather quiet place. I was glad when her father came in.

Saturday, October 20, 1883

When I woke up I heard that girl singing like a thrush, and then her father telling her to make less noise and not wake Mr. Ferguson. The girl can cook. She gave us a fine breakfast. After that I went out for a walk and told them not to wait dinner for me if I was not back at noon. I went down to see Mrs. Jamieson, who was very much surprised to see me as she knew there were no whalers in the harbor, so I had to sit down and tell her all about my trip. At noon Mr. Jamieson came home and wanted me to stay until he came again at five o'clock, saying, "The wife and girls will keep you busy with questions until I get back for tea." What a lot of questions they asked, all about me, how I managed, if I had gone to see my people and about the trip. After tea I had to tell Mr. Jamieson about the things that had happened. At seven o'clock I said I would have to be going to Mr. Thatcher's. Miss Flora asked, "Aren't you going up to see Lady Ross? She is always asking about you. She calls you her Highland boy." She suggested going up there with me. We went a roundabout way, the girl hanging on to my arm. She said, "Many's the time I have thought of you since that last night in the Governor's garden." I told her that she must not get serious or think too much about me or I would not go walking with her. She said that she could not help it, and asked if I liked the Thatcher girl. I told her that I did not, and said the girl was to be pitied for not having a mother to guide her, that she was really clever even though a bit forward. I thought to myself that it was just a case of one girl jealous of another. Just before we went in to Lady Ross', Flora asked if I would take her home at ten.

The servant opened the door and said, "Come in, Miss Jamieson and bring your company." The old lady looked at me, fell on my shoulder and kissed me saying, "My boy, my Highland boy." I had to explain why I went to Mr. Thatcher's first, because of the letters I had to deliver. She said, "Well, you are my boy and you are not going back to his place again." I had to tell her all about my voyage to Boston, Liverpool, and back here again. At ten o'clock the old lady told me to see Flora home and be sure to return, as she would wait up for me. On the way home, the girl teased me until I promised to have a walk with her tomorrow night after supper. Lady Ross was waiting for me when I got back. Since meeting this dear old lady three years ago, she has been more than a mother to me, kept my feet from the crooked way, kept me clean, watched over me and lectured me sometimes when I went up to the barracks among the drunken hussies that went there. God bless her.

Sunday, October 21, 1883

After a late breakfast, Lady Ross and I talked about Philadelphia, Boston, and Liverpool. She asked if Liverpool was as busy as ever. "When I came out here ten years ago," she said, "what a busy place it was with all the steamboats and shipping going up and down the river." As we went down to church this morning, arm in arm, she asked, "Robert, do you remember

ST. HELENA AND ST. PAUL DE LOANDA, AFRICA

the graces those old Scotchmen would say before meals? They were good men, very sincere and brave, fearing nothing but God. My husband used to tell me of the big Highlanders in his regiment and how he could depend on them in India. They never failed him. My husband was a brave man, a real Highland gentleman. I must show you his medals, for I prize them very much."

There were a good many soldiers in church but I knew none of them. They were a new lot, an English regiment. The Highlanders that I had known had been sent home. The old lady and I stopped in at the Jamiesons' for lunch. Mr. Jamieson and I had our tongues loose and talked of many places and things. It has been twenty-two years since he came to St. Helena, and except for one trip sixteen years ago to the Cape, for the government, has never been off the island. In those days the people at the Cape were nearly all Dutch, so Mr. Jamieson did not like it and was glad to get back. Lady Ross asked Flora to come up to her house for tea. The girl told me a great deal about St. Helena and about a small village of about seven or eight houses on the other side of the island called Sandy Bay. The people there generally fish and bring their catches to Jamestown by boat. I said that I would like to see the place. Flora said that it was too far to walk, being about twelve miles across the island, but the old lady said, "You can get donkeys to take you. It would help pass the time until your ship gets in."

After tea Flora and I took a walk up the valley on the road to Longwood. We sat down and talked a long time about Scotland and America. Before it was too late, I took her home. She said, "Good night, Robert, my dear."

It was ten o'clock before I got back to Lady Ross' house, but she was up. She brought out a case containing fifteen medals that her husband had won, among them the Victoria Cross. It was the plainest of all but the most highly valued.

Monday, October 22, 1883

After breakfast I went down to the jetty to talk to Mr. Jamieson. He told me where to get a couple of donkeys and said that his daughter would be only too glad to accompany me to Sandy Bay. The girl was pleased and packed lunch. There were about thirty donkeys to choose from. I picked out two good ones and asked how much for all day. He said five shillings and wanted to send a boy with us, but Flora said she knew the way. Up the hill we went, along a rough road, but shady and fragrant. At the fifteen-hundred-foot level the scenery was great. We stopped at Napoleon's spring for a drink of clear, cool water and went on. The trees along the way were all bent and leaning the one way with the trade winds. The air was clear and fine. There was no dull time as we were laughing and joking all the way. The girl seemed to enjoy it and had no thought or care, and had me laughing at her telling of a party of girls that went donkey riding and how the donkeys got balky and would not start but finally ran away home. Our donkeys went along at a nice jog. We finally saw the village way down below, but coming to a nice stream, sat down for a rest and to eat lunch. We tied the donkeys and let them have a drink and some grass to eat.

After lunch we went down into the quaint little village. The houses were of thick stone with roses on trellises all over the doors and windows. They had small but nice gardens. It looked real homelike. The wind blows on shore at all times. I noticed that their boats were good and strong.

We had a fine time and got back to town at dusk. The donkeys were tired and the girl too. After leaving the donkeys at the stable and taking Flora home, I went back to Lady Ross' house for supper and told her about the fine time we had. The old lady told me how as a girl of thirteen in Scotland, she used to get on the sheltie's back and go through the Black Glen and up and over the mountains and through the heather, some fifty-five years ago.

Tuesday, October 23, 1883

After lunch I strolled down to the Governor's garden. Two old sailors sitting on a bench hailed me, and then a third came along. All three were beach combers, getting old and gray but

ST. HELENA AND ST. PAUL DE LOANDA, AFRICA 225

apparently active. They had been here for three or four weeks with never a chance to get away. Asking the oldest man for a sea yarn, he finally consented and told a yarn of old ships and the Western Ocean. I listened to him, and if I was not mistaken he told a true story of himself. I asked him if he had any money and he said, "No," so I gave him five shillings and told him his story was worth it. I shook hands with him and was not ashamed to. When I got home, I wrote the story down for future use. Lady Ross gave me pen and ink and a lot of large sheets of paper and said, "Go on, Robert, my boy. Take your time." I had to stop and think sometimes but finally got the yarn all written down, the dear old lady sitting near and patiently watching me.

Wednesday, October 24, 1883

After breakfast I went down and got report of a whaler to the eastward of the island, coming in. At four o'clock, the good old whaler *Kathleen* came to anchor. I was glad to see her. When Captain Howland and Sam Hazzard came on shore, I met them at the landing. The captain shook hands with me. Sam, with a broad grin on his face, hugged me like a woman. The captain kept me busy answering questions as we walked uptown together: how I made out; when I got to Boston; where I was stopping. I told him and he was astonished to learn I was stopping with Lady Ross. I told him that she came from the same place in Scotland as I did.

We went up to see Mr. Thatcher. When he saw me, he asked, "Where in the world have you been?" He whistled after I had answered him.

The captain said, "Bob, you look very well, better dressed and younger. You don't have to get on board until morning but we are going to sail at nine o'clock."

I went up the street to Lady Ross' to tell her that my ship had come in and that we were going to sail in the morning. While I was having tea with her, in came Flora Jamieson who said, "Father said your ship came in to land a sick man and I thought maybe you had gone on board." The old lady asked her if she was tired after her donkey ride. She said she felt stiff yesterday but was all right now. The girl and I went out for a walk and had a long talk in the garden. She hung tight on to me. It was late when I took her home. I knew she felt bad but I could not help it, and I told her that I would bid her good-bye in the morning.

Thursday, October 25, 1883

Lady Ross kissed me as she bid me good-bye. I felt bad at parting from her, this dear old lady who has been so kind to me, God bless her. I stopped at the Jamiesons to say good-bye to them but did not see Flora. It was eight o'clock when I got on board. All hands seemed glad to see me, each one shaking hands with me. My boat's crew hung around me and my harpooner hugged me in real Portuguese style. I went below to change my clothes. My stateroom seemed very small compared to the after cabin on the *Daylight* and the stateroom on the *Dun Robin Castle*. For all that, I feel more at home here and more contented. Talking to Mr. Gifford after dinner, he thought that I had made fine time. When I told him about the Irishman, he laughed and said he would have liked to have seen me with my dander up, punishing him. "You should have put him in irons," he said.

The captain called me aft. Once more I had to give all the details of the trip and tell him how Mr. Aiken had offered me the captaincy of the *Daylight* or another ship still larger. "Why didn't you take it, Bob?" the captain asked. I said that my word was to get back as soon as I could and I never break my word. The captain said, "Bob, you are a man all through and I won't forget you." I told him how kind the owner had been to me; what a fine time I had at his home; how much money he had given me; and that he had bought my passage ticket.

"Well, Bob, I am glad that I gave you the chance to take the boat to Boston," the captain said. "Watch the ship tonight. We are heading for St. Pauls. Keep your eyes open and do not shorten sail. Tell Sam to keep her on her course."

It was my first watch on deck tonight and I certainly had to answer a lot of questions from the boys.

Saturday, October 27, 1883

We had not finished breakfast this morning when the familiar and welcome sound rang out from aloft, "Thar she blows, thar she blo-o-ows, sperm whales!" About a mile off on the port bow was a school sporting and playing. Mr. Gifford got fast first. Then Mr. Hazzard and Mr. Roderick both got fast. A whale came alongside of me. When I told Fernand to drive his irons into it, he let go both irons. "Down sail and mast," I ordered. "Roll them both up. Keep clear of the line. Pass sail and mast aft." The whale sounded. When the line slacked, we knew he was coming up. I yelled to Fernand to get aft, take the steering oar and lay me on top of the whale. "Haul line. Pull me right on top," I said as I lanced the whale again and again. "Stern all, stern all." We had to get back in a hurry when the whale began to thrash the water. Clots of blood came out of his spout holes. He kicked and went into a flurry, rolling over and over. Each of the four boats brought a whale to the ship.

Monday, October 29, 1883

Last night, after we finished boiling oil, the captain said, "Bob, I think you brought luck with you." After breakfast today we called all hands to stow down oil in the hold, 130 barrels, a good catch. I sighted whales in the afternoon from masthead. All four boats gave chase but we lost sight of them. We are nearing the African coast, for there are lots of birds to be seen. Though we sighted humpbacks, we did not lower for them.

FROM ST. PAUL DE LOANDA TO THE SOUTH ATLANTIC WHALING GROUNDS

Whaling fleet in harbor—St. Paul de Loanda—Unhealthy convict town—Ague and coast fever—Buying jewelry—Ferguson gives dinner to his crew—Monkey stew—Naked natives—Chain gangs—Drunken harpooner in a fight—Sailors dead from fever—Sick mate—Leaving port—"Bloody Dan" Gifford—Whales again—Rough and cold weather—Stove boat—Big whale takes out three tubs of line—Money spent.

CHAPTER XII

FROM ST. PAUL DE LOANDA TO THE SOUTH ATLANTIC WHALING GROUNDS

Tuesday, October 30, 1883
At masthead, this morning, I could see the mountains away back from the coast, where there was a light fog. As we came closer to the land, we could see the houses of St. Paul de Loanda. The captain came on deck and ordered all hands from aloft. Standing in the fore chains taking soundings and getting twenty fathoms, we ran into a harbor, past two or three forts, swinging the lead all the time. On the land side were the forts, and on the sea side the sand dunes. With a good land breeze, we ran in until we had ten fathoms of water, then we shackled a range of chain to the anchor and let it go in seven fathoms.

After getting the overhead boat down for a shore boat, the captain went on shore. There was quite a fleet of whalers lying here: the *Sea Ranger, Petrel, Andrew Hicks, Merlin, Pioneer, Stafford, Greyhound, Hercules, Alice Knowles, Mattapoisett, Milton, Niagara, Falcon, Eliza Adams, Jerry Perry*, and a merchant brig.

There was no wharf or pier, only a sandy beach. The water was as smooth as a mill pond. The low-lying city looked fine from masthead, quite a large place with some fine big buildings. There were lots of small boats in the harbor. The men gammed some of the whalers tonight but no one got on shore from our ship.

The mate told me that we were not going to ship any oil home from here, so there was very little to do. We got some water, but very little of that, for it was very poor and besides we have a lot of water left on board. They say that this is a very sickly place, lots of ague and coast fever. Several of the ships have already lost some men, big strong men who were carried off in one day. I think the big strong men go first, especially the men who drink the heaviest. Rum, wine, and gin are very cheap here. The men think liquor keeps the fever away, but it is the worst thing they can take.

Wednesday, October 31, 1883

This is a convict town. The population is made up of ticket-of-leave men, murderers, and thieves, all convicts from Portugal with numbers on their sleeves. I took my boat crew on shore today and invited them to have dinner with me. Fernand, my harpooner, and I were going for a walk, so I told the other men to meet us at a certain place at one o'clock. We walked through the business section and noticed there were some large hardware shops and some fine jewelry stores. Fernand wanted to buy a ring for his girl, and seeing some that he liked in one of the windows, went in to look them over. There was as fine an assortment as I have ever seen and very reasonable. He picked out a fine ring with five brilliant stones. The clerk showed us some bracelets of unique pattern, from which I selected a pair of rare beauties, and picked out two fine rings, asking him for the price of the lot. He told me one pound in English money. Any one of the rings was worth what I paid for the lot. I paid cash and for Fernand's ring too. When we left the shop, Fernand told me that the clerk was a convict. He was a young man, pleasant and polite, and wore a badge on his arm with his number on it.

We met the boys and went into a fine large restaurant which was airy and clean. As no English was spoken there, I asked Fernand to order a good dinner for the six of us. When served up, everything was hot with chili peppers that burned your mouth but tasted good. Lots of wine was put on the table but Fernand ordered coffee for me. The dinner and the coffee were good. I asked if the stew was chicken but was told that it was monkey. Well, it ate fine but perhaps it was just as well that I didn't know anything about it until I was all through. I don't know how many bottles of wine my boys drank, but it did not go to their heads. These Portuguese use wine in place of tea and coffee but they do not get drunk. When I paid the bill, all it cost was six shillings in English money besides the shilling I gave the waiter for giving me a monkey stew.

After dinner we went to a hardware store for pocket knives. We each bought one, paying a shilling each for a very fine Eng-

SOUTH ATLANTIC WHALING GROUNDS

lish knife. "Look, Mr. Bob, look," Fernand said, calling my attention to a young woman, as black as ebony, who had just walked into the store. There stood a young negro girl, as straight as an arrow and as naked as the day she was born. No one paid any attention to her as it was a common occurrence.

The police are all soldiers. Each carries a musket and a bugle. They need the bugle to call for aid, for they are a wretched-looking lot of men. Fernand wanted to stop on shore tonight. Before going back to the ship, I told him to keep an eye on the boys, not to get into any trouble and to keep out of the night air.

Thursday, November 1, 1883

Some of the streets in this old city were covered with sand eighteen inches thick that was blown in here some years ago during a sand storm. The convicts were digging it out, using wheelbarrows. There were twenty wheelbarrows to a gang and four soldiers to guard each. Each convict had a large iron ball chained to his leg. When a wheelbarrow was wheeled, the ball was put into the wheelbarrow. The soldiers were heavily armed with guns, swords, and pistols. I heard they had rather a hard time of it as they got only two meals a day and then mostly beans.

The stores in town are very good. You can buy almost anything, clothing, shoes, musical instruments, and so on. The merchants are all ex-convicts, quite polite and not a bit surly. An inspector keeps tab on them, making his rounds regularly. If anything goes wrong, back they go to the fort. There is a large stone fort at the back of the town with walls about forty feet high. It is about two blocks square with guards on each of the four walls night and day.

Mr. Gifford is very sick. As most of the men and officers are on shore, only Sam Hazzard and myself were left to look out for things. The captain came on board this afternoon and complained of one of our harpooners, Grinnell, raising the devil and getting drunk as a beast. It seems that he licked two policemen. After taking their guns from them, he chased them. Some woman hid him. The captain told Mr. Gifford to send for him tomorrow and get him back on board.

Friday, November 2, 1883

Yesterday four men from the whaling fleet died, and today two more. The harbor is crowded with ships. Three small coasting steamers and the mail boat from Lisbon came in today. The mail boat brought some more convicts, heavily ironed and well guarded.

Mr. Gifford called me today and asked if I thought I could get Grinnell back on board. I said that I would take my boat's crew with me to find him. Mr. Gifford said that if he did not come civilly, to knock him on the head, but to get him here somehow.

When we got on shore, I left one man in charge of the boat and cautioned him not to leave it until we got back. We went uptown, and after telling the boys what I wanted, we started to look for Grinnell. Before long, we met Captain Howland who asked if we were looking for our man, and he gave me five shillings to give the boys. After a while, Frank and Fernand came to me saying they believed they had found him. We went to this house and knocked on the door. We found him with his head banged up and blood all over his face. I spoke to him, wiped the blood from his face, arranged the bandage on his head, and fixed him up the best I could. When I told him to come along and get on board with me, he said, "Not a damned step." I sat down to reason with him and he said, "You praying parson, do you think you can force me to go on board?" I said, "If you don't come peaceful, I'll take you even if I have to carry you." Fernand told him that he had better come along and not make any trouble. Grinnell yelled, "Who the hell are you talking to, you damned Portuguese?" and made a pass at him. I stepped in front and said, "Don't raise your hand again." Grinnell made a pass at me. I was that angry that I swung for his jaw with all my force. He rolled over on the floor. Still mad, I kicked him in the ribs. He never moved. The woman screamed. I told Fernand to stop her cries and told Frank to fetch some water. After we soused him, he came to and went along quietly with Fernand on one side of him and me on the other. When we got to the ship, Mr. Gifford was there but too sick to say much, only telling me to fix up Grinnell. After I took the bandage off his head and clipped his hair, you could see an awful gap where someone had hit him with a club or a musket. As I stitched the edges of the skin together, he cried like a baby. Someone brought him a glass of gin. Then he went down into the steerage and fell into a sound sleep.

The captain told Sam and me to see that all of the crew were on board by nine o'clock tomorrow morning, because Mr. Gifford was too sick to attend to it. After I put the captain on shore and got back to the ship, Mr. Gifford asked me to copy some writing on the slate in his book. He had quite a few remarks to make about what had happened in the last few days. I thought it a grand thing to get to write in a ship's log.

My boys wanted to go on shore to spend the five shillings the captain had given them. I told them they could go if they would see that everyone got back and not too late.

Saturday, November 3, 1883

At seven o'clock the captain came on board from the *Milton*. Everything being ready, we headed offshore. This is a sickly port and we are lucky to be getting out of it. Mr. Gifford has not eaten anything and feels badly. When I went down to copy his writing in the log, I asked if he could not eat a light custard made with plenty of eggs. The steward having a slight touch of fever, I thought I would make it for him. It was the first thing he had eaten in four days and he felt better for it. Some of the men are not so well, but Grinnell's head is better.

Tonight we gammed the *Milton*. She had sailed out of St. Paul de Loanda with us. Two of her men died of the fever. The *Falcon* lost three men, the *Niagara* two, the *Sea Ranger* three, the *Stafford* two, and the *Pioneer* one. Ours was the only ship that had not lost a man.

Tuesday, November 6, 1883

Sunday afternoon Sam Hazzard got a whale. This afternoon all hands were stowing down oil in the lower fore hold, forty barrels all told.

Mr. Gifford is feeling a great deal better. He was up on deck a short while today. Whenever we go in to the coast, he always gets the fever and has to take quinine all the time. Out at sea he is a strong, healthy man. I made him another custard which he enjoyed. We broke out water, flour, bread, beef, pork, beans, vinegar, and rice. We spoke to the whaler *Bertha*. They have not seen a whale since leaving St. Paul de Loanda and reported that all the whalers are working south.

Thursday, November 8, 1883

Close hauled and with every stitch of canvas on her, the old *Kathleen* was heading west and by south with fine strong trades. Mr. Gifford, feeling much better, was up on deck almost all day. A very large four-masted ship, a Dundee clipper, went past carrying stunsails and a cloud of snow-white canvas. She was lead colored with painted ports. Her figurehead, a queen of the seas with a trident in her hand, was painted white and gold. What a beautiful sight for the eye of a sailor!

We have had lots of chicken to eat since we left the coast, but there are not many left. We have all of the pigs that we got when in port but have got rid of the ducks. A large ship was racing a barque, both going northwest, and it was nip and tuck. The men have been talking and wondering why all the ships went into St. Paul instead of St. Helena. Maybe that unhealthy place will be a lesson to some of them.

Friday, November 9, 1883

Mr. Gifford is feeling nearly well again and has quit taking quinine. He said to me, "Bob, we are bound south again for right whales. I'll be mighty glad to get into cool weather again for I always feel better then. When a fellow gets on the coast of Africa in those low-lying lands, he's likely to get the fever and anybody that gets it once will be likely to get it again."

This rough officer, known throughout the whaling fleet as "Bloody Dan" Gifford, for the first six months out from home never had a civil word for me. He used to call me a packet rat and worse, but for the last three years he has made amends and favored me many times. I have never had any dread of him for I thoroughly understood my business. I would be sorry to see him leave the ship. Thanks to my early training, I can hold my own with any of the men, whose good will I have from the captain down to the cabin boy. I dread nothing for I have always lived a clean life and have never flinched at any danger.

While Mr. Gifford was sick, I took observations on the sun for the captain. He asked me today how I managed so well on my trip to Boston with nobody to take a sight. I told him that I took the sights myself and showed the big Norwegian how to take the time by the chronometer; that we passed to the westward of Bermuda and when I raised the lights on the coast, I knew that my sights must have been right; and that I followed the Gulf Stream and had a good slant of wind all the way to Boston. I also told the captain that the barque *Daylight* was a good sailer, for she passed several ships and would have passed more had her bottom been clean. The captain then said, "Bob, call Grinnell for me." He gave this fellow a talking to and told him a thing or two about his behavior in St. Paul.

Saturday, November 10, 1883

Fine weather, strong wind and heeled over with all sail set. After dinner I relieved Mr. Hazzard at masthead. A half hour later, I sung out at the top of my voice, "There she blows, there she blows, one lone whale!" All boats lowered away and pulled to windward. The whale sounded and came up between the mate and me, heading leisurely towards our boat. The mate came up behind so that both our boats got there at the same time. Mr. Gifford motioned me to keep to windward while he kept to leeward. Both harpooners darted their irons into the whale. It sounded deep and fast. The nose of each boat was pulled down level with the water and the whale lines were smoking around the loggerheads. The mate kept bawling to Sam Hazzard to come and bend on his line. The lines were entirely out of the large tubs and there was not much left in the smaller tubs. My, but that whale went down deep! When the line slacked, I was yelling at Fernand to get the mast and sail down and the mate was bawling the same to Grinnell and to get them out of the road, clear of the line. The mate went to his bow with Grinnell at the steering oar and I went forward while Fernand did my steering. We yelled to the boys, "Get your oars ready and out." Up came the whale close to the boats. What a noise it made, just like a locomotive blowing off steam! With a boat on each side of the whale, we lanced it again and again. Mr. Gifford shouted to me, "Bob, get it back of the fin. That's the vital place." Being a large whale, Mr. Gifford got out his bomb gun while I was lancing, shot a bomb into it and yelled, "Lay off, Bob, lay off and slack line. Stern all. Stern all." The whale made a dive. In about ten minutes it was up again and fell over on its side, breaking one of the oars of my boat. The whale was dead. As the ship ran alongside, we passed the lines up and got the fluke chains on.

Tuesday, November 13, 1883

It was four o'clock yesterday before we finished boiling oil from that whale. Being too hot to stow away, it had to lay over until today. It was a good day to work in the hold, not too hot, but it took us almost all day to stow down the eighty-nine barrels. It was a good catch. If luck will only hold with us for the next six months, we will be happy, for that will end this whaling expedition. We saw a large crate in the water but there was nothing in it. We lowered a boat and caught a lot of fish for supper that were swimming around it, and we took the crate for firewood.

Saturday, November 17, 1883

Yesterday, we hauled aback, lowered a boat and gammed the barque *Petrel*, Captain Cleghorn. They had caught two small whales since leaving St. Paul. We heard that the *Mars* got four that totaled 150 barrels of oil. The *Petrel* lost three men from fever and the *Mars* two while they were at St. Paul. We were lucky to get out of there clean and well.

While I was eating breakfast today, the lookout yelled, "There she blows!" All four boats sailed into a school of sperm whales, each getting fast. There was no trouble killing all four whales. They gave eighty-five barrels of oil all told.

Thursday, November 22, 1883

This day came in with a gale of wind from the west. Standing up on top of the house aft, we could not see far as the sea was getting too rough. Most of the men were down below for shelter, but ready for a call on short notice. The sea getting rougher all the time, all hands were called at four o'clock to take in all sail but the fore staysail, mizzen staysail and the goose-winged main topsail. The water was sloshing across the deck. Big seas kept coming over the lee rail, which was down in the water most of the time. Everything around the deck had to be lashed with extra lashings.

As Sam Hazzard relieved me, he said, "Bob, this is one fine night for us to think about." The man standing on lookout on top of the try works lashed himself in place. He was warned to be very careful; to report any light; to be sure to look to windward and to bawl out in time. I went below to a warm bunk, took off my oilskins and rubber boots, and placed them where I could grasp them in a hurry.

Friday, November 23, 1883

The gale was still blowing and the seas were so heavy that the deck was full of water. To relieve the sloshing back and forth across the deck as the ship rolled, we had to open the lee scupper doors. I went aloft at noon to see if there was anything in sight. The *Kathleen* was rolling and tossing about like a cork. I saw a large ship hove-to not very far away under lower main topsail and fore staysail. She seemed to be making bad weather of it. The captain and Mr. Gifford have been up on deck all day standing by the wheel in the shelter of the after house, keeping a sharp lookout. It was my turn from seven to eleven. When I called Sam Hazzard out at midnight, I told him not to lie there dreaming of his girl. He laughed and said he wished he was there right now.

Saturday, November 24, 1883

As the day dawned the gale moderated, but there was such a heavy sea that the ship was rolling something awful. To steady her, we set fore and main topsails, jib and two staysails, or else she would have rolled the masts out of her. The captain coming on deck said, "That's right, Bob, steady her." As the ship got some headway on her and wore round on the other tack, she lay better, heading more into the seas. By noon, the wind being a good deal lighter, and coming from the north, we got the topgallant sails on her. This helped to get her on her feet, but there was still an awful swell running from the west. Even with the slats on, the dishes would not stop on the table. At night we took in the fore and main topgallant sails and all of the light canvas, and headed south, still rolling badly but with the wind pretty well aft.

Sunday, November 25, 1883

At daylight we were hove-to under goose-winged main topsails. Later the sun was shining bright but the wind was still very strong. Tremendous seas were rolling from the west. Nothing was in sight but a few gonies. The cabin boy feels the cold down here. He was never in cold weather before in his life and could only stay on deck a short while. The captain got a thick woolen shirt and a heavy warm coat out of the slop chest to keep the boy warm. The night sky was very clear and full of stars. I gazed at the Southern Cross and was wondering at it.

Tuesday, November 27, 1883

Yesterday afternoon, after sighting a whale, we lowered two boats and gave chase. It was so rough that we could not see the whale, except when on top of a high sea. Mr. Gifford darted at it but did not get it. After that we could not get near it again.

SOUTH ATLANTIC WHALING GROUNDS

Today the sea has gone down considerable. All hands were feeling good and very anxious to get a right whale. Fernand was telling Mr. Gifford about me and the monkey stew at St. Paul. They all had a good laugh about it.

Thursday, November 29, 1883

This is the first good day we have had for a week, so the men are all cleaning up and washing their clothes. Since today is Thanksgiving, we are celebrating with the last of our chickens and all hands had a good feed. All the time that I can spare, I spend studying navigation from a book the captain loaned me. At four o'clock the mate said that we had better wash decks. I thought to myself that they had had enough washing to last the voyage, but orders is orders. At dog watch I held school for Frank and John.

Tuesday, December 4, 1883

This morning Sam Hazzard sighted right whales. We lowered three boats and gave chase. Mr. Gifford having trouble, called me to come and help him get an iron into the whale in case his iron should draw out. We pulled up and got the iron in. As the whale lay there like a log, Fernand and I changed places. When I got to the bow, I lanced it good. Mr. Gifford was very tired, having been working on the whale for the last two hours. He cried, "Lance it, Bob, lance it good. Be careful not to cut the line with your lance." The whale commenced to spout blood. Sam Hazzard had come up on the other side, but just as he began lancing, the whale went into a flurry. We backed away in time but Sam could not back away fast enough and got his boat stove in the bow. The mate told me to go and help bring the ship down here. When we got the fluke chains on, it was seven o'clock, nine hours from the time we sighted the whale. That was a grand fight.

Wednesday, December 12, 1883

Counting one whale cut in on the fourth and two more since, we got 115 barrels of oil more. We were lucky, for just when we had the last of the oil stowed down below today, a heavy blow came on. We had to furl the foresail and mainsail, double-reef the fore and main topsails, and get the boats on the upper cranes.

Now we are lying hove-to in a gale of wind and a very heavy swell. The clouds are very dark looking. There is nothing doing today and the only things in sight are a few albatross.

It was good to get down below tonight. The two parrots sang out, "Hello, Bob, hello, Bob." I had to speak to them to keep them quiet.

Monday, December 17, 1883

We saw three ships all bound northwest, one of which was a full-rigged ship looking as if the gale had struck her hard, for some of her spars had been carried away and some of her yards were dangling around the lower foremast. Her crew was busy clearing away the wreckage of spars and tangled rigging. As this is a place of sudden squalls, I suppose she was struck without warning.

While we were gamming the barque *Desdemona* of New Bedford today, someone raised whales. Both ships lowered boats, but after a hot chase we had to give up as the whales went to windward too fast.

Tuesday, December 18, 1883

After breakfast when I went aloft, I sighted a lone whale and sung out, "Get the boats ready." The whale was on the lee bow going slow. In about twenty minutes Mr. Gifford got fast. Down went the whale, sounding deep. The mate called me to hurry and bend on. I happened to be close to his boat and just got there in time to bend on my line before the last of the mate's line ran out. That whale took all of the line from the mate's tubs and a lot from my big tub. It must have taken out over a thousand yards of line. After coming up with a breach half out of water, away it went with a rush to windward as fast as it could go with all four boats trailing behind. We were hoping that the lines would hold until we got near. Everybody hauled line and hauled hard. The whale sounded again but not so deep this time. It came up with a roar and started to run around in a circle. We got up near and put two more irons into it, with Mr. Hazzard lancing on one side and the mate on the other. The whale was getting tired. The bows of three boats were close up with the men lancing away at it. Thick clots of blood came from its spout holes. When it ran around trying to bump the boats, we had to back off for a time. After going into a wild flurry, the whale kicked out and lay over on its side. The ship was close by but it was nine o'clock that night before we got the fluke chains on. After the all-day battle the hands were sent below, leaving only two men to keep watch until daylight.

Wednesday, December 19, 1883

Cutting in was started before breakfast. The head was hove in on deck about ten o'clock and all the body blubber by noon. After dinner the slabs of bone were cut from the gum and the head was let go overboard. Albatross were all around picking up the crumbs of blubber and whale meat. We got the try works going full blast. Tonight it was very dark and we had to keep the bug-light burning to see getting around the deck. Burning scraps in it gives enough light to see good all around. This was a good whale, for we got seventy barrels of oil from it.

Sunday, December 23, 1883

This day came in with fluky winds that came from all quarters. I went to masthead until breakfast, sighting four ships far distant heading north-northwest; a school of blackfish and a few finbacks spouting straight up in the air. I went down to sleep but about eleven o'clock a ringing voice cried, "There she blows!" I dressed in a hurry. We chased the whale for three hours in a choppy sea but finally lost sight of it and got back on board for dinner about three o'clock.

We were just through eating when a whale came up alongside the ship. There was a rush for the boats. The whale went down and came up far to the leeward. We sailed and pulled, all four boats racing in a choppy sea. Our boat got ahead of all of them. Fernand stood up but could not see the whale. The ship signaled that the whale was up and to the leeward. We pulled and sailed until nearly dark when the ship set her colors at the mizzen peak to come back on board. These lads of mine were laughing and joking all the way back to the ship. They are used to a long hard pull and do not mind it a bit. The captain said, "Bob, where did the whale go?" I said that he went toward the south pole and beat me to it. "The whale had to go some to keep ahead of you and your crack crew," the captain said. I replied, "Well, captain, when it comes to a pinch, these men of mine are there. They will pull every pound that they can and require very little urging. I don't believe there is another crew in the fleet that can beat them."

Captain Howland said, "I believe it, Bob. There's not a heavy man in the lot. You must be the heaviest. How much do you weigh?" I said that I weighed about 172 pounds for the last eight years.

Our position today was around Lat. 38° 11′ S. Long. 10° 03′ E.

Tuesday, December 25, 1883

Christmas day came in with rough weather and a choppy sea. We sighted a French barque, a full-rigged Danish brig, and a large ship, all bound northwest. They all had their colors flying in respect of the day. The brig outstepped the others because she was not so deep in the water.

We had roast pig for dinner. This was the last of our fresh meat. From now on we will have to fall back on the old reliable salt horse.

Seeing a whaler boiling oil this afternoon, we ran down, crossed her stern and spoke her. It was the *Alice Knowles*. Her captain said they got a good-sized whale yesterday.

After supper Sam Hazzard and I talked about the folks at home and wondered what they were doing. After we hove-to for the night, I went below, talked to my parrot, read my Bible, and turned in.

Thursday, December 27, 1883

There was nothing doing all day. After supper Sam Hazzard asked, "How much money did you draw in St. Paul, Bob?" I said that I had not drawn a cent this voyage as yet. "How is that? How about St. Helena?" he asked. I said that I did not need to spend any money there.

"Yet you stopped on shore for several nights."

"Yes, Sam, but it cost me nothing. You know I got fifteen dollars bounty for raising whales, three pounds for the boat race, five dollars for painting the name on the *Milton*, and five dollars from the gentleman for showing his daughter all over the ship. Some of this money I gave to my boat's crew."

Sam said, "Bob, I am ashamed to tell you but do you know how much I have drawn? Two hundred dollars. The captain cautioned me about drawing more, that's why I asked you to lend me a pound in St. Helena."

"I gave that to you, Sam. I don't want it back. I should think if you are going to get married when you get back to New England, you would save your money."

"I intend to get married when I get home. Will you be my best man? We have been the best of friends for the last four years and although we've been through some hard times, we've never had a quarrel."

FROM THE SOUTH ATLANTIC WHALING GROUNDS TO ST. HELENA

Tough job cutting in during a gale—Race for whales with boats from other whaleships—A general gam—Grace—Parrots—Mate's boat upset—Merchant ship hauls aback to watch killing of whales—Barque George Henry out of drinking water—Killing whales in rough seas—Knocked into hold by blanket piece—Blackfish—Beating other ships' boats to whales—Fight between finback and killer whales—Whaleboats chased by killer whales—Lance rammed down whale's throat—Boat smashed to kindling wood—Race between ships.

CHAPTER XIII

FROM THE SOUTH ATLANTIC WHALING GROUNDS TO ST. HELENA

Friday, December 28, 1883
Today a strong gale of wind made us run close-reefed. We were plunging along in a big swell when the lookout sung out, "There she blows!" At the first rising, our boat was close to the whale. Before it saw us, the harpooner got both irons into it although the seas were very rough. It sounded very deep and when it came up Mr. Gifford also got an iron in. The whale ran around in the rough sea making the spray fly. We pulled up on it and when close the men got drenched, but we lanced it good again and again. To cut in, we had to start the bug-light. It was rough work with the sea rolling down to leeward. All that kept her steady was the heavy strain on the cutting falls with the blanket pieces, which went flying across the deck in spite of the cross-deck tackle we had on. The deck was all slippery with grease. The blanket pieces went banging around as we tried to lower them into the blubber room. It was hard to stand on the cutting stage at all and a tough job all around in a gale of wind in the dark. At daylight all hands went below for hot coffee and some breakfast. We got eighty barrels of oil from this whale.

Friday, January 4, 1884
Last night the whaler *Desdemona* lowered a boat and gammed with us until ten o'clock. They have had good luck for some time back and have done well in catch. They have a fine lot of men on board, mostly Yanks.

Today we watched a finback whale going a blue streak as if there was something wrong with it. First it would dive and then jump half out of water. Then it would lie on top of the water slapping its tail and making a noise like a bomb gun. It lashed the water until it was white with foam. Finally it went off at great speed skimming the water like a dolphin. It must have been attacked by killer whales but too far off for us to see what really was the matter.

Monday, January 7, 1884

Just after dinner the cry rang out from masthead, "There she blows!" three times. It just so happened that there were four whaling ships in sight getting ready for a general gam, but this cry nipped it in the bud. The orders came thick and fast to get the boats ready and away lively as the other ships were doing the same. Our crew tumbled into the boats like rats, no time to get shoes or hats. As I shoved away from the ship, some of my men had hard-tack in their hands. We up mast and sail with Mr. Gifford yelling to us to get there first as it was "all hog or none." I got a good start in a good breeze with Mr. Gifford close behind in his boat with its big lugsail. All my men were on the weather gunwale paddling for all they were worth.

The mate yelled out, "Get there, Bob. Get there. Paddle, you Portuguese devils, paddle hard!" My men just laughed and paddled the harder, gaining on all the other boats from the other ships. As we got near the school of whales, Mr. Gifford was three boat lengths astern of me. Our other boats were still farther behind with the other ships' boats, some twenty boats all told, racing for the whales. Mr. Gifford yelled, "Bob, go on, strike. Don't wait for me." I saw that I could beat the other ships' boats and sailed past the first whale and went on to the second to give Mr. Gifford a chance at the first one.

"Get up, Fernand," I cried. "Wait until I put you on top of it, then give him both irons." He did and clear up to the sockets. Mr. Gifford struck at the same time into the first whale. Away each of us went for a Nantucket sleigh ride. Mr. Gifford and I were lancing both, sometimes one and sometimes the other, whichever one was handy. Some of the boats from the other ships got fast too. Our second mate's boat came along and he did not go after a third whale because he thought it was mine, but when Mr. Gifford yelled to him to get it, then he got fast. Mr. Gifford killed his whale right away with a bomb gun to save time, because our whale lines were all foul of one another. The ship came along and got the fluke chains on all three whales.

WHALING GROUNDS TO ST. HELENA

The captain told me to go and help Mr. Hazzard bring his whale to the ship, pointing out where Sam was. It was five o'clock before Sam and I brought the fourth whale alongside.

Tuesday, January 8, 1884

It was after breakfast when the last of the whales was cut in. One watch went below to rest until noon and it was my watch at the try works. The men at masthead reported three of the other ships boiling oil.

Captain Howland said, "Bob, that crew of yours certainly can pull some." The mate spoke up and said, "Yes, and Bob's boat can sail some too. He beat me with my big lugsail."

"Mr. Gifford, wasn't it you who struck the first whale?" the captain asked.

"I don't know," Mr. Gifford said. "Bob was three boat lengths ahead of me, passing the whale I got and getting the second one. He cut out the mate of the *Falcon* by a boat length."

"Bob, you have good control over that crew of yours," the captain said. I replied, "I always treat them the way I would like to be treated myself. They are all good men and always do their best. When I ask them to do something, I never have to speak twice."

Some of the ships are gamming, but I hope they don't come here as we are too busy boiling oil to bother.

Wednesday, January 9, 1884

Finished boiling oil at noon but as it was too hot to stow down below, the casks were lashed on deck. At four o'clock, the captains of the whaling ships *Alice Knowles*, *Niger*, *Falcon*, and *Kathleen* went on board the *Milton* to have a gam. All the first mates went to the *Niger*, the second mates to the *Falcon*, and the other mates came to our ship. I was glad of it for I did not feel like going on another vessel. In the two hours they were here, there were more whales killed and more whaling done than in many a season. As they were all much older men, I held my tongue and had little to say except to answer questions.

Out of the last school of whales, the *Milton* got two, the *Niger* two, the *Falcon* one, and the *Alice Knowles* three. With the four that we got, it made a total of twelve whales out of the school, a very good catch.

Thursday, January 10, 1884

After breakfast Mr. Gifford, Grinnell, Fernand, and I went down in the main hold to stow away the casks of oil that Captain Howland, Sam Hazzard, and the cooper sent down to us from up on deck. All of the harpooners got busy getting their boats straightened out ready for another rush order. This evening we were standing around and laughing about some of the stories the officers heard on the other ships, and how the old mate of the *Milton* said, "That d—— fourth mate on the *Kathleen* seems to get there all the time ahead of everybody if he has a chance, but says nothing."

Friday, January 11, 1884

Not much doing today and everybody resting up. Standing by the main bitts, Mr. Gifford said, "Well, are you fellows catching the whales all over again?" Sam Hazzard replied, "The officers who came from the other ships caught enough whales to fill up and go home." The mate asked me if I had known any of them before. I replied that I had seen some of them in St. Helena but had never talked to them and so when they came on board, I had very little to say. Mr. Gifford said, "Bob, you gave them something to think about that doesn't often happen on a whaler."

I asked, "What was that?"

"I heard about you saying grace at supper the night the third mates came on board. They were noisy and rough in speech, but that pulled their tongues for them and gave them something to tell to the other ships when they gam. Not only that, but you got the pick of that school of whales. The most oil that any of the other ships got was seventy barrels by the *Alice Knowles* and we got a hundred and sixty-eight."

WHALING GROUNDS TO ST. HELENA

Frank Gomez and John Brown asked me to please give them a lesson, so I spent an hour and a half with them. These two boys are as good and faithful as they come.

Tonight, when I went below to my bunk, there was my parrot waiting, as usual, for me to come and have a talk. It always cocks its head on one side, looks at me and says, "Hello, chummie. Hello, hello, Bob." You would laugh to see this cunning little gray parrot climbing on to my shoulder, peering in my face, and looking around for peanuts or lump sugar.

Friday, January 18, 1884

The past week has been one of gales, high seas, and very rough weather. Our good old ship stood up fine as the pumps showed very little water. The sea has gone down considerable and the wind has moderated. At masthead I sighted two merchant barques and three full-rigged ships heading around the Cape of Good Hope. This afternoon we gammed the whaler *Desdemona*. She had a hard time in the gale and had two boats smashed in. They had seen a barque with her masts, yards, and bulwarks carried away and with her deck house stove in.

Monday, January 21, 1884

We sighted a lone whale, but after giving chase could not get close enough for a dart. Finally Mr. Conley of the *Milton* struck it, Mr. Gifford going to help. The whale was using its tail freely and making the water fly, keeping both boats busy for a while. Another whale made its appearance. Sam Hazzard got fast to that one and I went to help him. Both whales were towed alongside of the *Milton* where we finished cutting in by dark. The ships being mated, we divide the oil and the bone. On a big ship like the *Milton*, the work of cutting in and boiling oil is much easier, for there is lots of room on her deck. The *Milton* is twice the size of the *Kathleen* and has a flush deck. She is steady as a rock and being high out of water, her decks are always dry. These two whales gave each ship thirty-eight barrels of oil.

Thursday, January 24, 1884

At ten o'clock we heard Grinnell's melodious voice shouting, "There she blows, there she blows!" All four boats gave chase but could not get near. At four o'clock, sighting whales again, we chased until the mate got fast. By that time it was so dark that when Mr. Gifford pulled up to the whale to lance it, not being able to see very well, he cut the line with his lance or else it parted. The whale upset his boat completely but I got there in time to pick up his crew. We righted his boat and bailed her out. It was not easy to see the oars and the boat gear. However, we found most of it and got both tubs of line back in. It was after ten o'clock when we got back to the ship. We were soaking wet and hungry but the cook had some hot grub and plenty of hot coffee that did us a lot of good.

Sunday, January 27, 1884

After coming down from aloft for breakfast, I turned in for a nap. As I was just beginning to drop off, I heard the cry, "There she breaches, there she breaches, there she whitewaters. There she lobtails, there she lobtails." I heard the captain, "Where away?" and the answer loud and clear, "Four points off the lee bow." By the time I was on deck, the mate was shouting, "Steady ahead. Down with your wheel and meet her. Luff up a point. Get the boats ready."

The whales being quite a distance off, the captain went aloft. The whales breached again and again. Sometimes they lay on top pounding the water with their tails making a noise like a gun every time. The water became all white with foam.

When we were within half a mile of them, we rounded to, hauled aback and spread out all four boats near the place where the whales went down. Mr. Gifford being nearest, got fast first when the whales came up. Then Sam Hazzard got fast. The mate yelled for me to go on and help Sam. Sam's whale was sounding when I got there. When it did come up, I got a couple of good irons in. It started to run and pulled us around for an hour or more without our getting a chance to get near enough to lance it. The whale started on a line toward a large merchant ship

WHALING GROUNDS TO ST. HELENA 255

that was coming along. This ship hauled aback to watch the circus. I thought once that it would run into the ship, but we slacked line and the whale lay there, tired out. Sam and I came up, one on each side, and lanced it good. Thick blood came from its spout holes. All this time, the crew of the merchant ship were looking down on us and hollering at us, but we had no time to answer them. We were too busy keeping clear of the whale which was kicking, rolling, breaching, and pounding the water with its tail and running here and there. Every man was alert, ready at his oar to obey orders instantly. The men on the merchant ship looking on in wonder, cheered and waved their hats as the whale rolled over dead. Before they went on their way, they hoisted the Stars and Stripes and dipped the flag three times for us. Finished cutting in at midnight. Eighty barrels of oil and a lot of bone came from these two whales.

Tuesday, February 5, 1884

Pretty rough weather with a few snow flurries for several days, and the sea so heavy that there has been no lookout at masthead. We were sailing today in Lat. 38° 09′ S. Long. 10° 13′ E.

The man at masthead reported a barque signaling trouble, as her colors were upside down. She was the barque *George Henry*, a small barque belonging to Baltimore. She was out of fresh water and wanted to know if we could spare some. Captain Howland gave orders to give them twenty barrels of water in five-barrel casks and for the cooper to becket them for towing. After towing the casks over to their ship, they were going to hoist them with cant hooks but I told them that to do so would tear the chimes' to pieces and that it would be better to make some slings. When that was done, the water was emptied into their iron tanks and our casks returned to the *Kathleen*. The captain of the barque from Baltimore wanted to pay Captain Howland for the water but he would not hear of it. Our captain told him that we had one hundred more barrels of water on board, some of which would have to be emptied if we caught any more whales. This barque was bound to Natal in South East Africa from Pernambuco, Brazil. Captain Howland told me that her men had been on an allowance of one pint of water a day for a week. They would have stopped at Tristan da Cunha but it was blowing a gale and these islands are no place to be caught in during a blow. I well remember how when we were there I had to carry the sacks of potatoes on my shoulders out to the boat, slipping on the rocks and kelp, because it was too dangerous to bring the boat nearer shore, and that was during fine weather.

Sunday, February 10, 1884

This morning it blew so fierce that we took off all sail but the mizzen staysail alone. Then she lay much better and did not ship so much water. No one could go to masthead, for it was so rough it would have thrown him out of the hoops. Sometimes we had snow flurries or rain squalls. The men were huddled in shelter and all had their oilskins on to keep them warm and dry from the flying spray. The wind fairly howled and screamed through the rigging like someone in agony. The spray was driven across the deck so fast that it felt like hail.

Monday, February 11, 1884

After a rough night, the wind went down some this morning but it looked bad as it was veering around to all points of the compass. At noon a whale was sighted lying only a short distance from the ship. We wanted to lower the boats but the captain would not hear of it. Mr. Gifford asked leave to lower the larboard and waist boats with picked crews. The mate took me in his boat. We pulled right up to the whale. Although the water was very rough, both boats got irons into it. Down it went as fast as it could go, with lines smoking as they ran out around the loggerheads. Mr. Gifford told me to stay in the bow and help him while Grinnell took the steering oar. When the whale came up, it started off in a hurry pulling both boats behind it. The boats were jumping from wave to wave just like tin pans tied to a dog's tail, and throwing up clouds of spray from their bows. Just as we had hauled line close up to the whale ready to lance it, down it sounded again. The next time it came up, we hauled close and the mate lanced it good. Sam Hazzard was on the other side of the whale churning a lance into it. As it went into a flurry, we had to keep away a good distance on account of the rough water. Our boats were half full of water and the men were bailing it out with hats or anything. By the time the ship worked her way down to us, the whale was dead, but what a time we had getting on the fluke chains in that rough water.

After we had supper and some hot coffee, we had to start cutting in because the rough sea was making the whale pound the ship too hard. It was an awful job. We would cut off a blanket piece and hoist it, but before we could lower it into the blubber room it would go swinging across the deck in all directions. One had to be careful to stand in the clear of the swinging slab and watch for a chance to lower it into the hold. We had on cross-tackles but the deck was so greasy and slippery that a fellow could hardly keep his feet. We got the job done about two o'clock in the morning. The men were all tired and dirty and glad to turn in.

Afterwards the captain said that this was the toughest job of cutting in that he had ever experienced in all his days. Sam Hazzard said that he got knocked down once with a blanket piece. I said that once if I had not fallen down on deck in time as a big blanket piece swung, it would have smashed my ribs against the bulwarks. As it was, I got one good bump when the cross-deck tackle tore out of a big blanket piece and let me and the blubber down in the hold all of a sudden. It was lucky I was on the top side of the blubber. For all of our hard work, this whale gave us only twenty-eight barrels of oil and some whalebone.

Friday, February 22, 1884

We have been running north for about a week and are now having pleasant tropical weather. The blubber room has been given a thorough cleaning, as we seldom use it when sperm whaling, for then we leave the blubber lying on deck unless it is stormy weather with a gale blowing. At these times, we have to stow it down in the blubber room to keep it from sliding all around the deck.

The cook and one of the men forward had a scrap today. The cook hit William Wing on the head with a stick of wood. I parted them, sending the cook to the galley. I took William Wing, washed his head, clipped his hair and fixed him up. They fought over nothing, only an argument. One called the other a d—— fool.

Sunday, February 24, 1884

This afternoon the mates from the *Desdemona, Falcon,* and *Milton* came on board for a gam. Around four o'clock, noticing that the *Milton* went on the port tack, we put the *Kathleen* on the port tack too. Pretty soon masthead called out, "There she blows!" The *Milton* was lowering four boats. When the whales went down, we hauled aback and lowered four boats. Mr. Conley took my boat. I was to stay on board and be captain and work ship. Our four boats went out with mixed crews. When the boats were clear of the ship, I braced up sharp and followed them. I went to masthead and told the cooper to look out for the deck and the steering. The whales came up to windward. All sixteen boats from all four ships were trying to get there first, some sailing, some pulling, but the whales got farther and farther away each time they came up, finally getting out of sight and we called the boats back on board.

Monday, February 25, 1884

There were lots of tropic birds around the ship today, so tame that you could almost catch them with your hands. The water was covered with flying fish, the first to be seen for six months. They were being chased by dolphins, good big ones, three of which Sam Hazzard caught and they will make good eating. Mr. Gifford, Sam, and I were talking about Mr. Perry, the mate of the *Alice Knowles,* a Portuguese about fifty-five years old, who could not read or write. He is going to marry the school teacher at St. Helena. When they mentioned her name, I said

WHALING GROUNDS TO ST. HELENA

that I had met her. She was a fine-looking English woman about forty-five years old who had bought a controlling interest in the barque *Greyhound*. Sam said that the mate was going out as captain but that she would have to do the navigating. Sam could not understand why she should pick out a Portuguese. Mr. Gifford said, "Well, maybe it was take him or none."

Tuesday, February 26, 1884

At noon we hauled aback and lowered four boats after a school of blackfish. Mr. Hazzard and Mr. Gifford each got one. When we hoisted them on deck with a fish tackle, we saw that they were fine large ones from twenty-five to thirty hundredweight. We had just finished cutting them up when masthead called out, "There she blows, there she blows, sperm whales!" We gave chase and got one of them.

Friday, February 29, 1884

I went to masthead after dinner and right away sighted a sperm whale. Nearby was the *Eliza Adams* and another whaleship that were getting their boats ready to lower away. It was going to be a race for whoever was going to get that whale. There was a good strong breeze for the sails and I thought we had a good chance on equal terms. Mr. Gifford and Sam yelled to me just as I was leaving, "Bob, get there with that good Portuguese crew of yours and get fast before the other ships' boats. Let's see what you can do."

I told the boys to get their paddles and to dig deep but with no noise. I could feel the boat pick up and jump forward. I had Mr. Gifford four boat lengths behind when the whale came up between me and the mate of the *Eliza Adams*, but heading straight for me. I shouted, "Stand up, Fernand, and be sure to strike solid and not on the head. Paddle hard, boys. No noise. Paddle hard."

With the sail still carrying all she could, the boat rushed forward and the whale bumped right into it. Fernand sank an iron into it and the whale sounded. The other boats coming the other way nearly bumped into mine. We had to douse the smoking line with water. The nose of our boat was pulled down level with the water. Mr. Gifford came along and helped hold on to our boat and shouted, "Good boy, Bob. You beat the other boat to the whale." I replied, "By just two boat lengths."

Up came the whale and Mr. Gifford drove in both irons. With the mate on one side and me on the other, both lancing away for dear life, we soon had blood coming out of its spout holes.

The mate shouted, "Stern all, lively." We backed off eighty yards and let the whale kick it out, thrashing the water with its tail and snapping its teeth. When it rolled over, fin up, Captain Howland had the ship down close to us.

At supper the captain said to the mate, "I thought at first that a boat from the *Eliza Adams* had the whale and then I thought it was you, but later I saw that Bob got it. How was it, Gifford, that Bob got ahead of you with your big lugsail?"

Mr. Gifford replied, "Captain, there is no boat in the fleet that can beat him and his crew of Portuguese. They are just like a machine and paddle with no noise. When I told Bob to go on ahead, his boat fairly jumped ahead with these fellows stroking like Kanakas."

Captain Howland said, "Bob, you get five dollars for raising that whale for a bounty." I did not know about any bounty but thanked the captain and told him that I would give it to my boys, for they had worked hard and paddled like clockwork. "How long have you had that crew?" the mate asked. I told him ever since leaving Fayal and at that time they were the refusals of all the other boats.

Mr. Gifford said, "I was glad that you got ahead of the mate of the *Eliza Adams*, for he is one of the kind who knows it all. It did me a lot of good. I held my breath for a minute watching you head on to that whale. I was afraid you would shear off too soon and 'gallie' (scare) the whale, but you ran plumb into him before he saw you and then you sheared off. That was good work, Bob, for many an old whaleman would not have risked it for fear of those jaws. I was glad I let you go ahead of me."

Wednesday, March 5, 1884

A large ship came close and spoke to us, running alongside for a while. She was the *Pocahontas* of Norfolk, Virginia, bound to New York from Rangoon, India. When she rolled down, the barnacles on her copper sheathing could be plainly seen. In the afternoon a large school of porpoise, extending for miles around the ship as far as could be seen, were playing and sporting, even racing with us under our bow. What a lively thing a porpoise is!

Boats from two whalers were seen chasing whales to windward coming toward us. We lowered four boats and spread out to meet them but we never got near.

I have been fixing some clothes for the cabin boy, a pair of trousers and a blouse of white duck. When he got them he was as proud as a drum major. He is a very good boy and very grateful for anything done for him. He had a hard time on shore and had a fight of it to make a living, but he is on his feet now and knows enough to ship on any whaler.

Thursday, March 6, 1884

Seeing a finback breaching and making whitewater, we headed for it and saw that it was fighting about a dozen killer whales. The finback was worried and tried to shake them off. They hung on and bit the finback until the water was red with blood. Like wolves, the killers go in packs and will tackle anything that comes along in their way, no matter how big. We took a boat to go and get the whale but it sunk. The killers tackled our boat and we had a mighty lively time of it with lances and boat spade, keeping them off. We must have killed ten or twelve of them. When several came at the boat at one time, I was worried. As fast as they were killed they sank out of sight. We went back to the ship with our boat leaking badly where one of the killers banged his head into it. These killers have teeth about the size of your thumb. They are a species of whale with a large strong fin on their backs and run from eight to twenty feet in length. Like other whales, they have to come up to spout or breathe. They are very ferocious and dangerous because they run in large schools, mostly frequenting warm waters.

Sunday, March 9, 1884

Last night I went on board the *Milton* with our captain and stayed there all night. The *Milton, Falcon, Desdemona,* and *Kathleen* were all close together and their captains had a gam until two o'clock today when we all went back to our own ships. No sooner were we back when the man forward sung out, "There she blows!" The *Kathleen* was the nearest to the whales which were about a mile off to windward. Twelve boats from the other three ships were pulling like mad to get where we were. Up came the whales, one near Mr. Gifford. We all made a run for the school. The mate struck one first, then Mr. Hazzard got one, and as one came alongside of me, I got fast. The mate killed his whale very quickly but Sam and I had trouble with our whales which were of the fighting kind. Twice Sam's whale came near chewing up his boat. My whale bit off two oars and the only thing that kept it from cutting my boat in two was that I ran a lance down its throat. Mr. Gifford came up on the other side of my whale and lanced it through the heart. That done, the mate took his whale to the ship and I went to help Sam, whose whale had a kick at both ends, snapping its jaws and thrashing its tail. A boat belonging to one of the other ships was smashed to kindling wood. Luckily nobody was hurt and all of the boat gear was saved. This boat was so smashed to pieces that it was beyond repair and they had to let it go.

The *Milton* got one whale and the *Falcon* two, but all the *Desdemona* got was one boat chewed up.

Tuesday, March 11, 1884

Today we stowed down ninety-four barrels of oil in the main hold. The *Milton* got thirty-one barrels and the *Falcon* fifty-eight. All the ships are now heading towards the west and

St. Helena. The men are all anxious to go once more to the island where they have all had such good times.

Antonio, my tub oarsman, bought a ring in St. Paul de Loanda when the others were buying rings, but he has no girl. I told him that I would pay him what he paid for it. That pleased him. Fernand started to tease him saying, "You 'member we see him with big black girl in St. Pauls." Antonio spoke up and said, "Fernand, you one big lie on me. I no like."

Wednesday, March 12, 1884

As the ships *Hercules, Falcon, Milton, Niger,* and *Kathleen* were gamming this morning, a school of whales were sighted. Every ship put on all canvas and started racing for the school. The old *Milton* was in the lead. It is wonderful how this bluff-bowed ship can hang on to the wind against the ship *Niger* and beat her sailing. It was an exciting race up to sunset. Our ship was holding her own with the best of them. I was glad when my turn at masthead came so I could watch the race better. The race was given up at six o'clock, when all ships had a general gam. What an argument the men had about which ship was best! At ten o'clock the men went on board their own ships.

It was a beautiful moonlight night and a fine sight to see all five ships lying hove-to with the courses clewed up and the light sail furled. I did not feel like going to sleep so I wrote up my diary for two days back. Then I started to talk to Sam Hazzard. He said, "What's the matter? Don't you feel like sleeping? You had no sleep at all today. I'll bet you are thinking of some of the girls in St. Helena. My boatsteerer told me about the little blonde girl he saw you with in the Governor's garden, and how about the consul's daughter that Mr. Gifford was telling me about?"

"No, Sam, that sort of thing doesn't bother me much. Good night," I said.

AT ST. HELENA

Anchored at Jamestown—Letters—Sam Hazzard learns of his girl's marriage—Church services—Visiting old friends—Approaching wedding of school teacher and whaling man—Drunken harpooner—Invitation to the wedding—Liquor brought on board—Trouble with men on Mattapoisett—The Consul's daughter—Visitors from the mail steamer—The cabin boy's mother—Jealous girl—Dressed up for the wedding—Escorting Lady Ross—The wedding—The dance after the wedding—Songs and gifts at parting—Tender farewells—Off for home.

CHAPTER XIV

AT ST. HELENA

Friday, March 4, 1884

Yesterday all the ships were still headed east until breakfast time, then orders came to head west for St. Helena. The island came in sight this morning. The men were all anxious to get on shore and were cleaning and brushing their clothes, cutting one another's hair, shaving each other, and generally cleaning up. As we neared the harbor, the men got busy clewing up sails. The harbor was full of whale ships. We ran down among the fleet, rounded to and let go the anchor alongside of the *Morning Star* and the *Pioneer*.

Sam Hazzard took a boat down from overhead for a shore boat, landed the captain on shore, brought the mail back to the ship and asked me to distribute it. Nearly everyone got letters. I got two from Mother, bless her. Sam must have had seven or eight. Mr. Gifford asked Sam if he was going on shore again and he replied that he was *not*. From the way he said it, I knew something was wrong with him, that he was upset about something.

I told the mate that I would put him on shore. When I got back to the ship, there was the cabin boy all dressed up. He felt very bad that he had been too late to go with the mate, consequently I called my boat crew and told them to put the boy on shore. I told them, "Here's the five dollars I got for bounty when we raced for the whale against the other boats. Divide it among you." They were well pleased and said I was to call them any time and they would come for me, watch or no watch.

Sam was so grumpy that I knew something was wrong. However, it was not until after supper that he told me that the girl he was engaged to had married a widower with three children. He felt so bad that he could not tell me about it, just handed me the letter to read. It seems that her father had married again and that the new wife was the cause of it all, making things so unpleasant that Sam's girl was fairly driven to marrying the widower.

Saturday, March 15, 1884

Here we are lying at anchor at St. Helena with a pleasant breeze that keeps us cool. The captain wanted me to go on shore with him to see the American consul. I found out that an English battalion was stationed here now, together with a battalion of engineers from Mauritius. Meeting Mr. Jamieson down on the jetty, nothing would do but to go up to his house for tea. His wife and the girls were glad to see me. I told them where I had been since last seeing them and about St. Paul de Loanda where so many men from the whaling fleet had died of fever. They had heard about it from a barque that had put in to St. Helena to set five sick men on shore.

Mr. Jamieson said, "There is going to be a big wedding here soon. One of the whalers is going to marry the school teacher. She bought the controlling interest in the *Greyhound* and will do the navigating while her future husband acts as captain. After the school teacher gives up the school, my daughter Flora expects to get the position as assistant teacher." We talked and chatted until the warning gun boomed and then Flora walked down to the jetty with me.

Sunday, March 16, 1884

This morning Mr. Gifford said, "I suppose it's to the church for you. You'll not meet many men from the whaling fleet there. Do you know I have not been to church for fifteen years?" I replied that I had been taught to go to church and that I had never found it to my loss. At the church, Mrs. Jamieson took me by the arm to their pew in the front of the church where the rest of the family were seated. When church was over, Mr. Jamieson said, "Bob, you are a prisoner for the afternoon." After dinner we talked of Scotland until I thought I ought to go and see my old friend Lady Ross. Flora wanted to go with me. She took hold of my arm and walked along with me talking all the time. The way we went was no short cut.

She asked, "Do you remember our donkey ride to Sandy Bay? I was sore for two days but I never let on to you. What a good time we had!"

The old lady took one look at me and laughed and hugged and kissed me. After talking for a while, Lady Ross said, "Do you remember the school teacher you met at my house two years ago? She is going to be married to one of the whalemen, a Mr. Perry who is a Portuguese. He came up here with her one evening and seemed very polite, not a bit surly or rough." I replied, "I have heard of him but never saw him to my knowledge. He has the reputation of being a great whaling man." She told me they were going to be married in the church and have the wedding supper in the schoolroom. Afterwards there might be dancing at her house, but as it was rather small the dance would probably be in the schoolhouse.

At ten o'clock I took Flora home after promising my dear old friend to come back and stay the night. Lady Ross had a light lunch ready for me when I returned. We sat and talked about the coming wedding of the school teacher. She talked about weddings back in the Highlands of Scotland that lasted a week with dancing all the time and lots of whiskey, oat bread, and cheese. She asked me about my dancing. I told her that I liked to waltz and do the Spanish fandango. She said that when she was a girl she took dancing lessons and that her husband was a fine dancer.

Monday, March 17, 1884

The old lady let me sleep until nine o'clock. I thought that I'd better get on board. I soon had some men painting the ship and eight men painting the yards. In the afternoon we painted the jib boom, martingale, bowsprit, pawl, bitts, and windlass. I promised the mate that I would paint the name on the stern the next day.

After supper Mr. Conley, the mate of the *Milton*, came on board to see his chum, Mr. Gifford. They called me down to the cabin. They were much interested about Mr. Perry marrying the school teacher. Conley teased me about the Jamieson girls and the Thatcher girl.

After I had taken Conley to his boat, Mr. Gifford said, "Bob, how do you like Conley? He's a rough old boy and don't like the Portuguese. If Perry heard some of the conversation tonight he would not like it one bit, for he is very proud. He thinks himself a whaleman superior to any in the fleet. He plays all hog when he goes out for whales. Well, he's got the name for catching most of the oil. His boat was built to order, at least two feet longer and six inches wider than the regular boats. Now you have the smallest boat on our ship, yet with your sail and your boys paddling, I don't believe he would stand on an equal footing with you. I'm thinking there would be one wild Portuguese if he went racing against your crew."

Tuesday, March 18, 1884

I finished painting the lettering on the stern today as promised. Just as I was getting ready to go on shore with the watch, Grinnell came on board drunk as a fool. When the mate sent him below, he fell down in the steerage. His head was all bleeding when I went down to pick him up. After washing his head and fixing him up, I rolled him into his bunk.

As it was my turn on shore to stay all night if I wished to, I dressed up in my good clothes and went with the rest of the crew. Before I left them, I gave the boys a dollar apiece and told them not to buy whiskey or gin or I would not give them any more money. Fernand spoke up proudly and said, "This is not a Grinnell crowd. This is the bow boat crew."

Straight up the street I went to my old Highland friend Lady Ross, who said she was expecting me and that there was a lady there whom I had met before. I thought maybe it was Flora but to my surprise it was Miss Dickson, the school teacher who was going to marry the Portuguese, Mr. Perry.

Miss Dickson asked if I knew Mr. Perry but I replied that I had never had the pleasure of meeting him. "Well, Mr. Ferguson," she said, "I want you to meet him. The wedding is to be at the church. We are not sending out cards, just asking the people we want there. The reception is to be held in the schoolroom. I would like you to come down to my house now and meet Mr. Perry."

As I left to take the teacher home, Lady Ross said, "Do not

AT ST. HELENA

stop too long talking to Mr. Perry." We had only been in her house a few minutes when Mr. Perry came in. I found him very pleasant and quite a handsome man, looking not over fifty years old. He asked me to come to the church and to the reception after the ceremony. I did not stay long and went right back to Lady Ross. She told me that Flora was invited and supposed that I would be taking her. "No," I said, "Let her go with her father. I hope to escort you down to the church."

"Robert, I did not expect that you would do me the honor. You know I am too old to dance any more and will not stay for the dancing. There will be plenty of nice young folks to dance with. I know that Miss Dickson is forty-five years old but would pass anywhere for thirty-five. She is a nice dancer too."

Wednesday, March 19, 1884

On my way down the street to the jetty, I met Captain Howland who gave me a large package to take on board for him. We finished painting odds and ends around the ship, the skylight, water butt, brace blocks, harness cask, galley, after house, wheel, boat davits, the head of the mizzen mast, crosstrees, crossjack yard, and deck buckets. Any lettering that was to be done I did myself.

After supper, when most of the men had gone on shore, who should come on board but Mr. Gifford, tired and hungry, so I got him something to eat which he said tasted good, just suited his palate. He lit his pipe and said, "Bob, it's a queer thing that you never enjoyed a good smoke of a pipe and the soothing comfort it brings. By the way, what did you ever do with the two gallons of French brandy that you got from the barrels that we picked up off the Western Islands." I replied, "I gave Mr. Young one quart to rub my back with the time the whale whacked me. Again, when we were cruising south of Good Hope and I had charge of the watch, I used to give my men a good drink when they were cold and before they went below to turn in."

"Bob, I heard from Mrs. Captain Potter that you are going to Perry's wedding." I nodded. He continued, "You always stop with Lady Ross when on shore, don't you? You certainly got in the good graces of the best people on the island. Mrs. Potter says that the old titled lady is wealthy and has the finest house on the island. She says that you are also well acquainted with the Port Officer's family."

I told the mate that these people were Scotch and came from the same locality where my folks came from and that we liked to talk about it.

Mr. Gifford said, "Conley and I saw you going into the church one Sunday morning with Lady Ross on your arm. Conley said to me, 'There goes your fourth mate and an old lady.'"

I replied that I always went to church when I could and that it was through the church that I met and became acquainted with all these people.

Mr. Gifford went on, "The captain says you are the only man on board who has not drawn any money. I've done pretty well as I've only drawn twenty-five dollars."

"I have not drawn any money because I did not need to," I replied. "I earned twenty dollars in bounties and twenty dollars in other ways. I have not needed to buy any clothes and what I have are good. I have spent some money but have never blown it in foolishly. These good friends of mine won't let me spend any money."

Thursday, March 20, 1884

When work was over for the day, I told the men to get ready to go on shore. Sam Hazzard seemed very anxious to go too, so I let him go in my place. Mr. Gifford said, "Bob, go over to the *Milton* and ask Mr. Conley to come over." I took four men and brought Mr. Conley back to the *Kathleen*, leaving John Brown and John Wing on board the *Milton*, which pleased them very much.

Conley is a big red-faced Vermont man, a rough hard customer, but for all that I like his bluff jolly way. His language is not very choice at times, especially if talking about some person he does not care for. Perry is one he does not like, and if

AT ST. HELENA

the conversation drifted that way I heard plenty. They tried to tease me about going to the wedding and told me that I'd have to get a high hat, a white vest, and a swallow tail coat, but I knew them too well and only laughed.

Conley said, "Bob, I have respect for you. Rough as I am, I notice that you don't use any cuss words. How about the men when they don't do right?"

I replied, "Cursing doesn't help matters."

Conley thought that cursing helped to let off steam when angry. Mr. Gifford piped up, "Bob gets just as much or more out of the men as any officer on board including myself. He gives them a square deal and you know, Conley, what a boat's crew he has." I took Conley back to the *Milton* and brought back my two men. Mr. Gifford told me later that Mr. Conley had taken quite a fancy to me.

Friday, March 21, 1884

Most of the men are on shore and there is nothing to do for all of the work is done. Grinnell was drunk again today. One of the men must have brought him a bottle of rum. The mate called all hands and asked if any of them did it. They all denied it. Then Mr. Gifford asked Grinnell but he would not tell where he got it. Mr. Gifford was angry and said that any man who brought liquor on board would be put in irons.

After having dinner with the mate, one of the men called me and said that a boatsteerer from the *Mattapoisett* brought the bottle of rum to Grinnell. I reported this to the mate and he was wild. He asked me to put him on board *that* ship. When I did, he raised Cain with their three boatsteerers. They looked threatening so I drew up close to Mr. Gifford who said, "If any of you men try to come on board the *Kathleen*, I'll throw you overboard."

The men talked back and started threatening but quit as soon as their mate, Mr. Bumpas, came up and gave them a lecture. He told them that they were lucky to get off so easy, for it was a serious offense for which they could be locked up. Grinnell had been third mate on the *Mattapoisett* before we took him on as harpooner.

After supper Mr. Gifford and I went on shore to see Mr. Thatcher. While the mate went out for a smoke with the consul, his daughter played the concertina and sang for me. When she stopped singing, she asked, "Why haven't you come up to see me? When you came in on the steamer and had to wait for your ship, you only stayed one night at our house. Have I done anything to offend you?" I told her, "No." "Then, Mr. Ferguson, I want you to come up soon and spend the evening with me, for I'll soon be going home to America. Papa would like you to help him in the consular office." But I would not make her any promise. Just then Mr. Gifford and her father came in. Mr. Thatcher asked me how I was getting along with his spitfire of a daughter. She said that I was not a bit sociable. As the mate and I were going down on board, I said, "That girl may be all right but I don't like her."

Mr. Gifford said, "Her old man is not very well and is anxious to have her married to some steady man. Both of them took a fancy to you ever since you showed the girl all over the ship. Mr. Thatcher asked me to bring you up here but to say nothing to you about it. If you meet her at Perry's wedding, treat her fair because she does not like the Portuguese."

Saturday, March 22, 1884

This morning some visitors came on board from the mail steamer that had stopped here to let some of the passengers visit the island. Mr. Gifford asked me to show them around the ship and explain things. I showed them the windlass and how it worked when heaving in the blubber; the two large pots in which we boil the oil; the cooler for cooling the oil; the cutting gear; how we hove in the big blanket pieces that weigh about two tons; and how we kept the oil in casks on deck until it had cooled down because the hot oil makes the casks shrink and leak. All this while, one of the young ladies was taking notes and said that she wanted to see inside the little boats that we used to catch the whales. I called some of the men to swing the

AT ST. HELENA

waist boat level with the rail. With a little assistance, she climbed into the boat and looked things over. I was afraid that she would cut herself feeling the keen edges of the lances. She wanted to know why we had two tubs of whale line. I had to be very particular because she was taking notes all the time, so I told her that the large tub had eighteen hundred and the small tub nine hundred feet of line, but that if the whale sounded very deep, that these two tubs were not enough and one of the other boats would have to come and bend on line, the whale sometimes taking out from thirty-five hundred to four thousand feet of line. When she asked about the loggerhead, I told her that sometimes the line went out so fast the loggerhead would catch on fire and the line would smoke, making it necessary to throw water on it to keep it from burning. She took notes of the length of the oars, especially the twenty-four-foot steering oar, so I had to explain the why and the use of it. I told her about the mast and the sail and how we had to get them out of the way when the whale was fast and be careful to keep all clear of the running whale line. Before I finished answering all her questions, the mail steamer whistled, calling all passengers back on board.

This fine-looking English girl thanked me for my courtesy, and her uncle, an old gray-haired man, shook hands with me and left a sovereign in my hand. Mr. Gifford told me that the girl's mother and father were both dead. She was some fine girl. She had such beautiful hands that I was afraid I would hurt her when I helped her in the whaleboat. I asked Mr. Gifford to let me have the honor of carrying our guests back to the mail steamer. On the way I said to the girl, "Now you have the pleasure of riding in a whaleboat and I will show you some of the advantages of the big steering oar."

I showed them how quickly a boat could be turned around using it, and with my good crew got them on board the steamer long before the rest of the tourists got there.

Sunday, March 23, 1884

The cabin boy came on board to see me today. I asked him if he had been paid yet. He said, "No." I asked him if he was going on to New Bedford. He said, "No, I would rather stay out here." I told him to see Captain Howland tomorrow, for he could get him a place on one of the other ships that would not be going home for a year or two. The poor boy's mother, so he told me, is drunk all the time that the whale ships are in here. I said, "I hope you don't touch liquor, for you have seen what it does to a person." Two years ago this boy was quite delicate, but he is strong and healthy now. He has always been very manly and obedient on board. When he returned on shore with the men, the tears rolled down his cheeks as he said good-bye. I feel for him, poor boy, as he has no one to guide him, and hope that God will watch over this fatherless boy and correct the mother's ways.

Monday, March 24, 1884

A large barque coming to anchor and looking familiar, I took the glasses and saw it was the Australian and New Zealand barque *Romance* that we first met in Fayal four years ago, and again two years ago, here. Sam Hazzard and I decided to go on board her and see if the same officers were there. We found them all and they were glad to see us. They planned to stop here three days to let the passengers see this historic island. As we left, they gave us a pressing invitation to visit them again.

Not waiting for supper, I went on shore to see Lady Ross, but did not get far when some one tapped me on the shoulder. It was Captain Potter's wife who said she would thank me very much if I would get a message out to Mr. Conley, that there were some passengers going on board with her tomorrow and that he was to get the ship cleaned up. I hailed a boat and delivered the message to Mr. Conley who afterwards put me on shore again.

Upon reaching Lady Ross' house, I explained why I was so late. The servant brought in cheese, bread, jelly, cake, and tea and then I felt better. Presently Flora came in and said she saw me talking to some woman on the street and then go back to the ship, consequently she did not expect to find me here. Flora said that her father and mother wanted me to come to their

AT ST. HELENA 277

place for dinner tomorrow. Thanking her, I said I would like to. Then we had some music. Lady Ross sang some Scotch songs that I had not heard since I was a boy. Flora sang some Scotch songs too. As I sat there listening, I thought that I was at home once more. The ladies started to talk about the coming wedding of the school teacher. Flora said that her mother was making the wedding dress for Miss Dickson and asked if I was going to the wedding. When I replied, she said, "I am so glad, for I never was at anything like this. I don't know how to dance either. Will you come to our house that night?"

"No," I answered, "I am going to the church with some other girl." She wanted to know who it was but I told her to wait and see. "Is it that Thatcher girl?" she asked. I would not tell her and told her that she would find out soon enough. After I took Flora home and was telling Lady Ross about it, she said, "You like to tease that girl. Well, when she sees you and me coming to the church, she will understand."

Tuesday, March 25, 1884

On my way down the street this morning, Fernand stopped me and said that the captain told him that if he saw me to tell me to come down to Mr. Thatcher's place. A little further along, I met Jim Lumbrie and his girl. Jim also said that the captain was looking for me. The captain was just leaving the consul's door as I came along. He asked if I had given Mrs. Potter's message to Mr. Conley. I told him that I had. "All right, Bob, come into the consul's office and help me with some papers." We worked until noon when the captain asked me to have dinner with him, but I replied that I was expected up at Mr. Jamieson's for dinner. "Bob, you seem to have gotten acquainted with some fine people on the island. You have not drawn any money all this voyage. Do you need any?"

"No, sir," I replied, "I don't need any."

When the Port Officer and I arrived at the house, they were waiting lunch for us. Mr. Jamieson had to leave after lunch but left me there with the women folks. His wife, who was busy sewing on the school teacher's wedding dress, said, "Bob, you are favored by being asked to the wedding. Only one of the captains is asked and that is on account of his wife, Mrs. Potter."

I said, "I am not anxious to go, except if they dance I will like it."

"The schoolhouse has a fine big room to dance in," Mrs. Jamieson said. "It's about one hundred feet long by seventy-five feet wide and has a splendid floor. There was a dance there five years ago when some friends of the Governor came from the Cape on a visit. They had a grand time and danced all night. They stayed a week and then went to Ascension Island, with a gunboat to carry them, being Government officials. Flora says that you are going to escort some girl to the wedding and won't tell her who it is. She is very anxious to find out." "Well, Mrs. Jamieson, she will see the night of the wedding." We must have talked about four hours when I said I must be getting back on board. I left them up to their eyes in work, sewing on all kinds of garments. Mrs. Jamieson is the only one on the island who has a sewing machine. Her people sent it to her. It is a wonderful thing. Everybody on the island came to see it and find out how it worked.

After supper, on the ship, I got my good clothes out, gave them a good brushing and smoothing, and polished up my buttoned gaiters.

Wednesday, March 26, 1884

This afternoon Sam and I paid another visit to our friends on the *Romance* as she is about to leave. They were anchored in a deep place and had ninety fathoms of chain out, so Sam and I each took our boat with its crew to help them up with the anchor. They were right glad of the help. These English sailors started with a chantey, "Good-bye, Fare You Well." We did not leave until we had helped them set all sail. It was a good four miles from our ship when we left her side. The Englishmen gave three rousing cheers for us Yankee whalers.

This evening, everybody but five men forward and myself aft was on shore. Mr. Gifford was on board the *Milton*. At ten o'clock I went over to her to bring the mate back, but Mr.

AT ST. HELENA 279

Conley asked me to stop on board. He was telling about two of Mr. Perry's old shipmates, both Portuguese, who had not been asked to the wedding and that they were feeling sore about it.

Thursday, March 27, 1884

A fine day and nothing to do. The men came straggling on board like a lot of lost sheep. Talking to Sam Hazzard, I said, "I miss visiting with the Highland soldiers who used to be here. I suppose we will soon be on the last leg of the voyage. Well, Sam, with your girl at home married and another girl here, I suppose you'll be in no hurry to leave." "No, Bob, I don't like this girl like the other but I am trying to forget," Sam said.

After supper I took a run on shore to see Lady Ross. When I got there, she had company, Miss Dickson and Mr. Perry. He turned to me and said, "I understand that you are going to escort Lady Ross down to the church." I told him that I considered it an honor to do so.

After they went, the old lady remarked, "Mr. Perry gives me a different impression of a Portuguese. I have always pictured them as rough and uncultured, but Mr. Perry is courteous and very much of a gentleman."

I asked Lady Ross if she had said anything to Flora about our going down to the church together, but she had not. At that very moment who should come in but Flora who said she had seen me coming up the street and came over because she wanted to see me. The two ladies started to gossip and all the talk between them was about the dresses Mrs. Jamieson was making; Miss Dickson and Mr. Perry; the wedding and the people who were going to be there. Flora said she would have to be going back home because the wedding clothes had to be finished tonight.

Of course I would be willing to walk home with her. The old lady urged me to come back and stay all night but I told her that I had to go on board and would see her tomorrow.

Instead of going home, Flora said, "Let's go to the Governor's garden and sit down." The benches were all occupied so we had to go up and sit on the rocks along the promenade. She coaxed and coaxed me to tell her who I was going to escort to the church. "Was it anybody she knew?" I would not tell her and said she would have to wait until the wedding tomorrow.

The warning gun sounded. The girl hung on to me and would hardly let me go. I just had to pull away from her.

Back on board, I heard Mr. Gifford saying, "What's up, Sam? Is all your money gone?" Sam replied, "It looks like the last night in port with all hands broke."

The mate said, "Bob, tomorrow is a busy day for some people. Tonight some of them are excited, especially that girl of Thatcher's. She is so worked up about the wedding, that she can hardly wait. She tried on several different dresses and asked her father and me which she looked the best in. She's a right good-looking girl, Bob, and she asked me if you were going to be there. I told her that you were a special guest. You dance, don't you? Well, she wanted to know if you did. Now, Bob, you give that girl a dance. She's really a good sort of girl and it would please her if you did."

Friday, March 28, 1884

It's a splendid sunshiny day that speaks well for the wedding. In honor of it all of the vessels of the whaling fleet have their flags hung out to the breeze. Two merchant ships and the old whaler *Greyhound* have flags from the jib boom to the fore masthead, from there to the main truck and mizzen topmast down to the spanker peak and taffrail. It looks like a real holiday. All of the Portuguese are dressed up in their best with all their jewelry, rings, earrings, and fob chains.

I thought that my blue clothes would be too heavy for dancing so I got out my white ducks, put on a white collar and my buttoned gaiters. Boatload after boatload of men were going ashore. When I arrived at Lady Ross' house, she said, "I thought you were not coming until after five o'clock." It was after five but her clock was twenty minutes slow. The old lady said, "Robert, I'd like you to wear this scarf." It was a very pretty silk tartan of the Ross clan, that had belonged to her husband. I thanked her, took off my necktie and put it around my neck instead.

This dear old titled lady and I started down the street arm in arm. It was crowded with people, many of whom we knew. Once in the church, we went well up in front in Lady Ross' pew, where we found Mrs. Captain Potter seated. We were quite near the Jamiesons who looked surprised to see me escorting the old lady.

The bride looked real pretty. Mr. Perry was well dressed, wearing a blue cutaway coat and white vest. He had a heavy watch chain across his breast and too many rings; too much jewelry, I thought.

After the wedding service, we went to the schoolhouse and had a feast. Lady Ross saying that she was fatigued and wanted to go home, I went with her. She insisted that I return for the dancing and afterwards come back to her house for a good rest and that I would find the door open.

The tables were being cleared away getting ready for the grand march when I got back. Couples were being paired off. I paired off with Flora although her mother told me that she could not dance. I said that Flora could walk and we did the grand march together. I asked her if her mother could dance, and she said, "Yes, she is a good dancer." As they were just forming ·for a square dance, I asked Mrs. Jamieson. She would not hear of it and had to be coaxed, so my first dance was with Flora's mother. The bride was dancing the lancers with a fellow who jumped around like a frog but who could not dance. My partner laughed and said, "That was the groomsman." The bride claimed the next dance with me. She was a very fair dancer. Next, they were forming for a fandango, a Spanish dance that is graceful and pretty when danced right. I asked Mr. Thatcher's daughter to dance it with me. She said that she had not danced at all yet and wanted to know where my partner was. I told her that I had taken Lady Ross home. Miss Thatcher astonished me as she was a light quick stepper and very graceful, whereupon I took her for a partner and danced the rest of the evening with her, with one exception, when Mrs. Captain Potter asked me to dance with her. Mrs. Potter said that her husband was not there and that she did not want to dance with a Portuguese. It was a mazurka. We went gliding over the floor. When I went back to get Miss Thatcher, she said, "Let these old women dance with their own men."

At three o'clock in the morning the dance was over. I thought it nothing but right to see this girl home after dancing with her most of the night. Not a thing would do but I must come in and have some refreshments. It did not take her long to put a fine lunch on the table. She was talking all the while about how she hated the Portuguese and how she had enjoyed the dancing. Her father heard us talking and came in to hear about the doings. She told him that she had danced most of the night with me and that I knew how to dance. I told him that I had learned from a Spanish girl down in Cuba. He asked me to stay until morning but I told him Lady Ross would be expecting me.

Saturday, March 29, 1884

The old lady let me sleep so long that it was ten o'clock before I got back on board. The water boat came along and we stowed down forty barrels of water. After dinner, when the starboard watch was going on shore, Grinnell begged the mate to let him go too and made all kinds of promises about not drinking. Mr. Gifford let him go and said to me, "There is a good man that money is a curse to. If he had none, he would be better off. He is a sensible man when sober, a good boat-steerer, a good harpooner, and reliable in a pinch. If he were not so good natured, I would not keep the big baby another day."

Captain Howland came over and wanted to know all about the wedding and the dance. He also asked about the old lady I escorted to the church. The captain told me that his sister-in-law had told him about having a good dance with me.

After Captain Howland left to go on shore again, the mate said that he had promised Mr. Thatcher's daughter to have me come up before the ship left. Most of the men have come on board with all their money spent, but not one drunk in the lot and all in good humor. It's a hard trial for an officer when

AT ST. HELENA

the crew comes on board drunk, in a quarrelsome mood ready to fight among themselves or with the officers. An officer must quell them the best he can and never show any fear of them.

Sunday, March 30, 1884

I stood on deck this fine Sunday morning listening to the bells tolling and the bugles sounding and thinking it is the last time I shall hear these pleasant sounds for some time. We are all ready to sail. The rigging is all down on deck and on the pins and the last of the fresh water is on board. Grinnell came on board half stewed, laughing and good natured, so no one said anything to him.

After dinner the mate told me to hurry and dress if I wanted to go on shore as the boat was waiting. Mr. Gifford and I went up to the Thatchers' to bid them good-bye. The American consul was pleased that we had come back, and his daughter asked us to stay for tea. When the two men lit their pipes and went into the garden for a smoke, the girl sat down beside me and said, "Mr. Ferguson, I like to dance with you. I don't like the Portuguese and as for dancing with them, they are too clumsy."

Asking her for some music, she got her concertina down and played some very pretty tunes on it. Then she sang "The Spanish Cavalier," which was all the rage at the time. I asked her if she played any Scotch tunes. It seems that she had learned a good many of them from one of the Scotch soldiers that used to be here. He could play the flute and was a grand singer, so she said. She played, "Flow Gently, Sweet Afton" and "Annie Laurie" for me.

Presently, I told her that I would have to be going as there were some folks that I would have to bid good-bye to. When I got up to go, I gave her the ring that I had bought from Antonio. She liked it very much and threw her arms around me saying, "Scotchman, I like you." Before I left, she sang, "Good-bye, God Bless and Keep You." What a fine voice she had for singing!

Shaking hands with her father, I went up the street to my

dear old friend Lady Ross, leaving Mr. Gifford sitting with the Thatchers. Lady Ross asked where I had been. When I told her that I had been at the consul's, she looked displeased and scolded me because I had danced with the Thatcher girl all the night of the wedding. I replied, "I had to dance with someone and the girl knew how to dance very well. Of course, I had to go up and bid her and her father good-bye."

Sitting in a dark corner of the room was Flora. I had not seen her before. She had very little to say and seemed to be rather angry with me.

Lady Ross said, "Robert, my boy, I hear that you will start to sail back to America tomorrow."

I said, "Yes, everything is all ready. We leave tomorrow as soon as the captain comes on board."

We chatted about the Highlands, the places we knew so well, the rugged rocks, the glens, the stormy islands, about Dun Donald and his place at the head of Loch Broom. The Dun Donalds were all great soldiers and could muster four thousand men. Later Dun Donald liberated one of the countries in South America where one of the finest of monuments has been erected to him. We spoke of many places where superstition is rife, of witches, and of the many ghost stories that were told about some of the Scotch places. Later the old lady played the piano and Flora sang "When Twilight Falls."

Flora asked how I liked the song. I replied, "So good that it feels like more." After we had some refreshments, I took the ring I had bought at St. Pauls and handed it to Lady Ross. When she opened the small box, she was astonished and said, "How pretty! Such a unique design, but I could not take it from you." I said, "I bought it especially for you. I want you to have it." Lady Ross said, "Robert, my boy, now I know that you used to think of me when in the wilds of Africa." "Of course I did," I replied. "It will be a long, long time before I forget you and all your kindness to me."

All the while the girl was looking at the ring and then at me. This dear old lady said, "Robert, that is a fine stone in the ring. You must have paid a good price for it." I said, "Not more

AT ST. HELENA

than you deserve." Whereupon she kissed me on both cheeks. Flora sang another song for me. It brought such a lump in my throat that I have never forgotten the words.

> I do not ask remembrance in your hours, busy and full,
> Bearing such gifts to others, rich in power for use and rule,
> Check not the current of your life that breaks joyous and strong,
> To hearken where some haunting memory speaks like a sad song.
> But when the dusk is creeping and the dew lies on the hill,
> When the first star is trembling through the blue, remote and still,
> When through the lilies steals a breath so faint, it thrills like pain
> And hushing into peace, day's long complaint, night falls again,
> Oh, then one moment be the present fled and that sweet summer
> That so strongly led in one our ways
> When I was yours in every pulse and thought, and you too seemed
> To give back something of the gifts I brought, or so I dreamed
> And know that as it then was with me, sweet is it still
> That a life's love is waiting at your feet whene'er you will.

Lady Ross said, "Good-bye, Robert. God bless and protect you," as we parted. It was hard to say good-bye but I had to take Flora home. As we went up the street she was very silent until I gave her the pair of bracelets that I bought for her in St. Pauls. She said, "Robert, you did think of me as well as of Lady Ross. You know the old lady is very fond of you." Flora would not go straight home but went around by Donkey Lane. We sat down and talked a long, long time. I thought I ought to go and say good-bye to her folks. "Wait a little while, Robert," she said, and then threw both arms around my neck. After a while we went up to her house. Mr. and Mrs. Jamieson were sorry that I was going away and hoped that I would come back soon.

The signal gun boomed. Flora said that as this was the last time, she would walk to the jetty with me. She liked the bracelets and said that there were none in St. Helena like them. I told her to look inside and she would find her initials. When it came time to part, it was hard, for she clung to me and kissed me over and over.

"Kiss me good-bye, Robert dear, for I may never see you again." I was so glad that she never shed a tear. I could not bear to look back as I got into the boat waiting to take me on board.

I turned in, but not to sleep, thinking of Flora and all these kind people who have been so good to me.

FROM ST. HELENA TO NEW BEDFORD

Stowaways—Boxing matches—Passed Ascension Island—More whales—Stove boat—Hot weather—Long talk with Captain—The doldrums—Trade winds—Tough fight with whale—Dancing—Music—Bad storms—Approaching New Bedford—Home again.

CHAPTER XV

FROM ST. HELENA TO NEW BEDFORD

Monday, March 31, 1884

This day came in with fine weather and a nice breeze. All hands were on board, none missing. After breakfast we took the gaskets off all sail and hove up the anchor with a chantey "Fare You Well My Bonnie Young Gal, We're Homeward Bound." When the captain came on board and the shore boat was hoisted up on the crane aft, we hoisted sails and got the anchor inboard and lashed. As we stood across the bow of the *Morning Star*, we fouled her jib boom but did no damage. With all sail on, we headed northwest on the last leg of our voyage, bound back for New Bedford. Some are happy and longing to get there and home. Somehow, I was not so anxious to leave St. Helena, for I was quite contented there.

On this return voyage we have only three mates, Mr. Gifford, Mr. Hazzard, and myself. Mr. Roderick is no longer second mate as Sam takes his place and I am third instead of fourth mate. Mr. Roderick is on the ship but not as an officer.

It was my first watch below tonight, but I did not feel like turning in. I was awake for a long time thinking of all that had happened on the island. It must be that thoughts of the girls back there disturbed me.

Tuesday, April 1, 1884. Lat. 14° 45′ S., Long. 7° 02′ W.

All sail set and booming along in fine weather and strong trade winds. One man was forward standing on the fore royal and a harpooner at the main royal on lookout, the hoops or rings having been taken down.

We found four stowaways on board, one from the *Niger*; one from the *Mermaid*; and two from the *Falcon*; all Portuguese and apparently good fellows. There were two barques sailing along the same as we were but they slowly dropped astern. A ship with stunsails from her royal yards down to the water's edge gained on us although she was very deep in the water. Running before the wind makes the sun feel hot. When washing down the deck at four o'clock, the men were having a good time throwing water on one another to cool off.

Wednesday, April 2, 1884. Lat. 13° 25′ S. Long. 8° 22′ W.

Since Sam Hazzard, my roommate, is in the starboard watch and I am in the port watch, the only time we get together is during dog watch. He was wondering what I was going to do after I got back to New Bedford, and spoke of what good times we could have together at Westport Point, Adamsville, and Central Village. He does not feel so bad about his girl any more, only at times. He can't understand how she came to marry, only having another six months to wait until he got back. I told him that he would just have to wait and see what was wrong with her.

I wonder what my folks would say if they saw the Bible they gave me, all stained inside and outside with whale oil. I hear Grinnell down in the steerage singing all alone. He must be feeling good again.

Thursday, April 3, 1884. Lat. 12° 30′ S. Long. 9° 39′ W.

We caught a few fish today, small but good. Frank Gomez and John Brown asked for a lesson. I taught them for two hours and found them both doing very well. When Sam Hazzard came down from aloft to the vise bench where I was holding school, he asked John Brown if he was going to open a school when he got home to Brava. John pulls midship oar in Sam's boat and is a very good fellow.

Saturday, April 5, 1884. Lat. 10° 08′ S. Long. 12° 36′ W.

It's fine tropical weather and very warm, with nothing for the men to do but play cards, checkers, or dominoes. Some of them are boxing with a set of small gloves that they got from a soldier in St. Helena. These Portuguese know very little about boxing, although one big black fellow thinks he does. Sam and I put on the gloves but I saw right away that Sam knew nothing about it. All hands were watching the boxing matches. Even the captain and Mr. Gifford seemed to enjoy them. The

TO NEW BEDFORD

big black fellow put on the gloves with Sam, who knocked him over the main hatch. For the next bout, Grinnell took on a big Portuguese, but he was only a play toy for the big harpooner.

Grinnell called, "Bob, come on and let's see how you use your dukes." Mr. Gifford said, "Bob, try him out," and Fernand said, "Go on, Mr. Bob."

"Well, to please you boys, I will," I replied, putting on the gloves.

At first, I felt out Grinnell to see what he knew about boxing. I soon saw that I had him if I wanted to, for he had no defense and his guard was poor. I let him hit me once or twice to make him think he had me. Grinnell began to rough-house it. Waiting for an opening, I yelled, "Look out," swinging for his jaw as I said it. It staggered him and must have made him mad because he commenced trying to let me have it hard. I made a feint. He fell for it and bit. Stepping quickly to one side as he was coming toward me, I caught him on the side of the jaw. Running into it, the blow had double force. Like a bullock, the big harpooner crumpled up for the count. The Portuguese, looking on, thought Grinnell was dead, for it was ten minutes before he came to. When the mate threw a bucket of water on him, he came to and heard the canaries singing. Captain Howland took the gloves and threw them overboard.

"Why did you hit him so hard?" the captain wanted to know.

"Grinnell tried to get rough with me," I replied. "I thought I'd give him a lesson that would do him for some time."

Mr. Gifford told the captain that it was not the first time that I had to do the trick, as it had happened before in St. Paul de Loanda when the only way to fetch him on board was to knock him out. I guess the captain had not heard about that.

"I had no idea that you could give such a beating to a big fellow like Grinnell," the captain said. "You are usually so quiet, Bob, never swear and go to church."

"Captain, my first voyage was a rough one and I picked up some knowledge of how to defend myself. Later, on some of the western ocean packets, it was often fight or not eat. In those days, it was mostly brute strength, no science, just power.

There was no clean fighting, always foul play and dirty work. I found out there was science to fighting, so the first chance I got, I learned how from a good boxer."

"I should have liked to have had you with me one night in St. Helena," the mate said. "There was a street fight where men rolled all over the street, biting, kicking, gouging, with their faces all covered with blood."

Sunday, April 6, 1884. Lat. 8° 01′ S. Long. 15° 02′ W.

Today we sighted Ascension Island, a coaling station for the British Government. It's a lonely barren rock in the middle of the Atlantic Ocean. There is no fresh water, and what they have is condensed from sea water. There are several large government storehouses, two or three small steam launches and a gunboat. The place is so barren that no vegetables can be grown, and it is very hot. There are lots of fish and plenty of birds' eggs can be had for the gathering.

Monday, April 7, 1884

This morning, when Fernand was on lookout on the royal yard, he sighted a school of sperm whales and hollered, "There she blows!" All four boats chased the school. It was a scramble with all of the boats trying to get fast and the whales running every which way. Finally, all of the boats got fast and the lines became all foul of one another. The whales were luckily not sounding very deep. One boat got badly stove but killed its whale. My whale was trying to switch its tail around so that it would strike the boat, or if not that, it was coming at us with its jaws wide open. All the while, I was standing up in the bow with the lance in my hand, watching for a chance to lance it. When the fun and the circus was all over, we had four whales. They gave us fifty-eight barrels of oil which we stored down below. Tonight the Southern Cross hung low on the horizon. It will soon be out of view as we work northward.

Friday, April 11, 1884. Lat. 3° 41′ S. Long. 21° 51′ W.

While the days are very hot as we near the equator, the nights

TO NEW BEDFORD

are quite pleasant for sleeping. To try to sleep below in the daytime is stifling and choking.

The men and some of the officers are making fancy mats to take home. Blackfish were spouting and playing all around the ship, heading to the north by the thousands.

It was a fine night on a glassy sea. The men sang Portuguese songs accompanied by a leaky accordion. It was squeaky music that was pumped out of it, for these men were used to strumming the strings of a guitar.

Saturday, April 12, 1884

After breakfast everybody was looking for a shady place, as the sun was blazing hot. There was not much to do except wet the casks in the hold, pump ship, and wash the deck. The weather is so close, mucky, and oppressive a man gets so languid that he does not feel like exerting himself. The men relieve the heaviness and liven themselves up by throwing water on each other. The wind has been so light that we have had but little steerage way and have to keep hauling the yards and braces around to catch the slightest bit of favorable breeze.

Monday, April 14, 1884. Lat. 0° 16′ S. Long. 20° 00′ W.

The sea is as smooth as a mill pond. The wind is very light and it is awful hot. It helps some by keeping the decks wet. In hot weather like this, we are careful to keep the casks in the hold wet. John Brown, who is planning to go back to his home in Brava, was talking to me about the money he had drawn on this voyage. In the four years all he has drawn has been about twenty-five dollars, and that paid for the clothes he got out of the slop chest. He has been careful, taken good care of his clothes and has not squandered his money foolishly. He will get a good little sum of money if the New Bedford sharks don't get it.

Friday, April 18, 1884. Lat. 2° 46′ N. Long. 29° 14′ W.

Last night, about midnight, a good strong blow from the north made us call all hands to shorten sail. They were on the jump for half an hour. Although the blow was severe and the lightning blinding, we did not lose a stitch of canvas.

Today it was warm again and everybody was having fun sousing each other with buckets of salt water. The breeze was strong and we went hopping along with both royals set.

The ship was surrounded for a while by porpoises. As they come playing under the bow of the ship, we dart at them from the martingale. When we get one, it is hauled up on deck by a block fastened to the fore rigging.

Sunday, April 20, 1884. Lat. 4° 40' N. Long. 33° 15' W.

Not much doing today. Sam Hazzard was asking me about the girls in St. Helena. "Who was that girl you were dancing with at Perry's wedding?" he asked. I replied, "I danced with Mr. Thatcher's daughter most of the time as she was the best dancer there. I also danced with Captain Potter's wife."

Sam said, "No, the one I mean was a blonde, the one Jim Lumbrie saw you with in the Governor's garden."

"She was one of the Port Officer's daughters," I replied. "She is a peach but does not dance. When she saw me dancing with the Thatcher girl, she went home with her mother."

"Well, Bob, you must have had a good time and were mighty lucky to get acquainted with such nice people."

"Sam, do you remember how the first night you got in to the island, you and Tommy Wilson slept in a boarding house where the bedbugs woke you up in the middle of the night? Tommy said you were so disgusted that you got up and went out."

Sam said, "I thought Tommy knew all about the town because he had been there so often, but all he knew was the grog shops."

Tuesday, April 22, 1884. Lat. 7° 43' N. Long. 38° 02' W.

During my watch last night from seven to eleven, Captain Howland asked if I was glad and in a hurry to get home. I said that I was quite contented on the *Kathleen*. "Have you a girl in Philadelphia?" he asked.

TO NEW BEDFORD

"No, sir," I replied.

"I know that you had a girl in St. Helena to go around with. My sister-in-law, Mrs. Potter, told me that you danced all night with the American consul's daughter."

"Yes, sir, I did, for she was the best dancer and there were not many to dance with."

"Have you a girl in New Bedford, Bob?"

"No, sir, I only got in to New Bedford the day before we sailed."

"Now, Bob, watch close and call Mr. Gifford if you need him." That was the longest talk I had with Captain Howland concerning my affairs in all the four years of this voyage.

Today all sails are drawing and pulling right along with a fine breeze. Everybody seems cheerful and jolly now that we have struck the steady trade winds again. I can hear Grinnell, a happy-go-lucky, good-natured fellow and with a dandy voice, singing, "Oh, I Am The Wind The Seamen Love."

Many times I have heard this song while in the trades. The first time I heard it was when I was second mate of the fruiting schooner *Samuel Wackerell* of Boston, when we met the trade winds after weathering out a hurricane. This was the same blow that wrecked the Government ship *Huron* when all hands lost their lives. Five hundred and sixty people on board the steamer *Richmond* were all lost in that same storm off Cape Charles and Cape Henlopen. We rode it out with the loss of but one man. Our bulwarks were stove and one boat was lost. I always think, whenever I hear that song, how glad we were to pull through that hurricane.

Sunday, April 27, 1884. Lat. 13° 55' N. Long. 46° 06' W.

Fine weather and good trades holding. All sail set and bowling right along with every stitch of canvas pulling. The old *Kathleen* looks fine and clean and new, all but the sails, which are getting old. All of the men are talking of what they are going to do when they get to New Bedford. I'm thinking that some of these air castles will fade away like a dream.

Thursday, May 1, 1884. Lat. 17° 31' N. Long. 51° 10' W.

After breakfast I turned in for a few hours' nap, but awoke about eleven o'clock with Sam Hazzard shaking me and singing, "The Whale Fish is Blowing." As I got on my feet, I heard Grinnell's cheerful voice crying, "There she blows, there she blows, and there she whitewaters!" in a ringing tone that would have awakened all hands had they been asleep. Three boats rushed towards the school of whales with all the sail they could carry and the men paddling like clockwork. All you could hear was the swish of the water running past. Up came the whales close to the boats. Mr. Hazzard struck one. Although I was in the middle of the school, my boat was the last one to get fast.

As the mate got a lance into his whale, his boat got stove. Mr. Hazzard helped the mate kill his whale, for, like mine, it was one of the fighting kind. It rushed head on to their boat and rolled over so that it snapped at them with its jaws.

My whale lay on top flinging its tail around like a flail. All of a sudden it came for me head on. I got a lance into its eye. That made the whale mad. It breached out of the water half way, then stuck its head down, swinging its tail in all directions.

While this was going on, the mate and Sam got their whale killed and towed to the ship.

I had to watch my chance to pull up to the whale very cautiously. It came running towards the boat with its jaws wide spread, towering over us like enormous pincers. Quick as a thought, I rammed the lance down its throat. It backed away with my boat following. At last a lance was rammed into a vital spot. About the time Sam Hazzard came back to help, the whale rolled over dead, fin up.

It was three o'clock when we got the fluke chains on. We had just started to cut in when someone raised sperm whales. That meant that the boats, lines, and gear had to be straightened out before we could give chase. We followed the whales until dusk when we had to give up and return to the ship.

Saturday, May 3, 1884. Lat. 19° 33' N. Long. 53° 49' W.

The cooper repaired the mate's boat and fixed a split plank

TO NEW BEDFORD

in mine, caused by my whale bumping into it when we were towing it to the ship. Twenty-nine barrels of oil were stowed down in the main hold.

Sam Hazzard was telling me of his first voyage to New Zealand in the whaling barque *Abram Barker* as cabin boy. It was a four-year voyage that seemed like a lifetime to him. He said that it was squally all the time down there and often it was so rough they had to run for shelter. We talked together for a couple of hours about places we had been. Sam thought that this voyage had been a pleasant one for both of us. I said that I had no regrets and felt satisfied because I had seen so many things that I had hankered for.

Thursday, May 8, 1884. Lat. 26° 40' N. Long. 61° 10' W.

This day came in with fine weather, a gentle breeze and all sail set. Some of the men are making bird cages and door mats. Others are lying around on the warm deck waiting for a strong breeze to hurry us on, for they all seem anxious to get home.

This evening after dog watch, the men forward were dancing the fandango, "sham a reta, sham a rosa, ans a posa, dona posa" and kept it up and kept it up to the accompaniment of a fiddle, a guitar, and a squeaky accordion. All the time the men kept snapping their fingers like castanets. Finally Grinnell mixed with them and danced the hornpipe. For a big heavy man, he is quite spry and light on his feet. We have a happy, laughing, good-natured lot of men. About ten o'clock a large school of sperm whales came all around the ship and quite close, spouting almost alongside. For more than fifteen minutes they were running right along with us. It was such a clear night that you could see them spout quite plainly. What an opportunity if it had been daylight, but we did not disturb them!

Sunday, May 11, 1884. Lat. 29° 40' N. Long. 64° 59' W.

The wind hauled around to the east this afternoon, with heavy squalls, keeping us busy handling the sails, bracing around the yards, first on one tack and then on the other. After sunset it cleared up, but it felt close and sultry as if there was something brewing.

Sam Hazzard said to me, "Listen to that big porpoise singing. You would not think he ever thought of a girl, only of a bottle of rum." It was Grinnell singing in a very good voice "We Parted From Each Other."

Friday, May 16, 1884. Lat. 35° 30' N. Long. 70° 00' W.

A heavy sea kept coming over the rail in tons. The decks were flooded. The wind went down and all hands were called to put sail on her again. This evening there were more squalls, and all hands had to be called to shorten sail. The lightning was sharp and continuous. The thunder was deafening, with one loud roll booming right on top of the other. It was a nasty, dirty, stormy night with a rough sea rising and slopping over the deck.

Saturday, May 17, 1884

This day came in with a gale of wind from the east, a regular snorter. Around noon, we tried to get some sail on her but had to take it right off again, as the wind hauled to the north. We had to heave to under short canvas and wear ship every two hours.

Monday, May 19, 1884. Lat. 38° 36' N. Long. 69° 26' W.

This morning the weather was fine with a light breeze from the north. The sea had gone down considerable. We knocked down the try works and threw the bricks overboard. The try pots were turned upside down and lashed. The bulwarks and decks were scrubbed. We wet the hold with the hose. The water ways and mast were given another coat of paint.

I put a lock and new hinges on my sea chest. Some of the things that my chest would not hold, like the mats, I had to bundle up. All the while my parrot sat on the edge of my bunk saying, "What's the matter, chummie?" and kept looking at me with his head cocked on one side, kind of knowingly.

Sam's parrot was laughing just like a human being. His bird

TO NEW BEDFORD

does not like the cage, so it has had the freedom of our stateroom ever since he got it at Kabinda before it had any feathers on it. Now, when we are going home, it has to get used to a cage.

Wednesday, May 21, 1884

Yesterday we had more squalls with heavy rain and the wind shifting around to all points of the compass. We are trying to work in towards the coast and are looking for land. I notice that the water has changed color from a dark blue to a lighter dull gray.

Today started with a drizzly rain and stayed foggy all forenoon. When it cleared up in the afternoon, masthead reported land in sight. We made out Gay Head and No Man's Land, firing our signal gun two or three times.

We began to take soundings and had forty-five fathoms of water; at four o'clock, twenty-eight fathoms; and twenty-five fathoms at eleven o'clock.

The boys have been on deck most of the day but don't seem a bit anxious to go below to rest.

Noticing that there was no riding light out, and knowing it was dangerous to be without one, I told Sam about it, so he hunted one up and hung it in the fore rigging. In St. Paul de Loanda or St. Helena we did not need it, but here I guess Sam and the mate forgot. We hauled aback until daylight.

NEW BEDFORD

Thursday, May 22, 1884

This morning the hands did not need any calling. All of them were out on deck on the alert for the first break of day. About five o'clock we were standing in towards the harbor with all sail set.

As we passed the Dumpling and the Point, we started to take in the light sails, clewed fore and main royals, took down the flying jib and gaff topsail, jib, fore and main and mizzen staysails, topmast staysails, and clewed fore and main topgallant sails.

When we passed Clark's Point, Fort Phoenix, and the coal pockets, we clewed fore and main sails, let go topsail halyards, headed into Merle's wharf and got the hawsers fast.

The men are all happy, shaking hands with friends or busy getting their things on shore. Sam and I stood on the deck watching them. We had nobody to meet us.

Then Sam yelled, "Give that calf more rope!" He was yelling to Grinnell who was down below in the steerage singing at the top of his voice,

> Through rip and swell
> In gust or fog
> In sleet or rain
> Let me go back to the watch
> On the old *Kathleen* again.

APPENDIX

Glossary of Nautical and Whaling Terms

The "Lay," Catch, or Wages

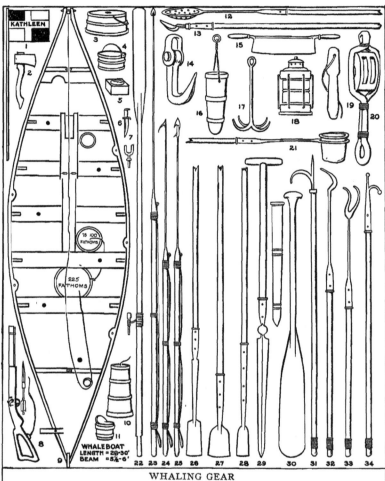

WHALING GEAR

1. Waif, or Boat Flag.
2. Boat Hatchet
3. Boat Keg
4. Boat Bucket
5. Boat Compass
6. Boat Knife
7. Oar-Lock
8. Bomb-lance and gun
9. Whaleboat
10. Lantern Keg.
11. Boat Piggin.
12. Skimmer.
13. Blubber Pike
14. Blubber Hook
15. Mincing Knife.
16. Case Bucket
17. Boat Grapnel.
18. Lantern
19. Toggle.
20. Cutting Tackle
21. Bailer.
22. Steering Oar
23. Lance.
24. Single Flued Iron.
25. "Temple's Iron" (or toggle-harpoon)
26. Bone Spade
27. Cutting Spade
28. Head Spade
29. Boarding Knife and Sheath.
30. Boat Paddle
31. Gaff.
32. Fire Pike
33. Blubber Fork
34. Boat Hook.

GLOSSARY OF NAUTICAL AND WHALING TERMS

ABACK—A square sail is aback when its forward surface is pressed back by the wind.

ABAFT—Toward or at the stern of a vessel.

ABEAM—Opposite the side of a ship, at right angles to the keel.

ABOUT, TO GO—To change direction when tacking.

AFT—Near, toward, or in the stern.

AFTER—Applied to any object in or near the stern.

ALEE—On or toward the side away from the wind.

ASTERN—Behind the ship, or in the direction of the stern.

ATHWART—Across, from side to side.

AWEATHER—On or toward the side toward the wind.

BACK THE SAILS OR YARDS, TO—To expose the forward surface of the sails or yards to the wind.

BACKSTAYS—Ropes from masthead slanting aft to sides of ship.

BAILER—Vessel used to dip water out of a boat, also a dipper used on a whaler to transfer boiling oil from the try pot to the cooler.

BAILING THE CASE OF A WHALE—To dip the spermaceti out of the whale's head.

BANDS—Strips of canvas sewn on a sail to prevent it from splitting.

BARE POLES—A ship is under bare poles when no sail is set.

BATTENS—Strips of wood nailed on the slings of a yard, or over the tarpaulins of hatchways to keep out storm water.

BEAM, ABAFT THE—Said of the wind blowing between "on the beam" and the stern.

BEAM, BEFORE THE—Said of the wind blowing between "on the beam" and the bow.

BEAM, ON THE—Said of the wind blowing at a right angle with the keel.

BEAT TO WINDWARD, TO—To make progress to windward by tacking.

BECKET—A short strap or rope for suspending a yard until wanted.

BECKET, TO—To secure by beckets, for instance, to tie a raft of casks together when rafting water.

BELAY, TO—To fasten a rope to a pin or cleat.

BELAY THERE—Stop, quit. that's enough.

BELAYING PIN—Stout pins around which ropes are turned to secure them.

BEND ON, TO—To tie on, as for instance, to tie on another length of whale line to one nearly run out.

BIGHT—A loop, the double part of a rope when bent.

BITT, TO—To secure the cable around the bitts.

BITTS—Fixed, vertical timbers or iron castings for securing the cables or hawsers.

BLANKET PIECES—The large pieces of blubber first stripped from a whale.

BLOCK AND BLOCK—Said of two blocks when pulled together as tight as they can go, also called chock-a-block.
BLOCKS—Sets of sheaves or pulleys for multiplying rope power.
BLOW, TO—Said of whales when spouting.
BLUBBER—The fatty covering of whales and the like, which is rendered into oil.
BLUBBER-HOOK—The heavy hook used for stripping the blanket pieces from the whale and hoisting them on board.
BLUBBER-ROOM—A room between decks where the blubber is stored until it can be cut into smaller pieces.
BLUFF-BOWED—Having a broad and flat bow.
BOATSWAIN—The officer in charge of the rigging and whose duty it is to summon the crew.
BODY-OIL—Whale oil made from the blubber from the body.
BOLSTERS—Pieces of wood or stuffed canvas placed on the lower trestletrees to prevent the rigging from chafing.
BOLT-ROPES—Rope sewed around the edges of the sails.
BONNET—An additional piece of canvas laced to the foot of a sail in moderate weather to hold more wind.
BOOMS—Large poles used to extend the studding sails, spanker, or jib.
BOOM ALONG, TO—To sail at a lively clip or under full sail.
BOW—The forward part of a vessel, the prow.
BOWLINES—Ropes made fast to the leeches or sides of the sails to pull them forward.
BOWSPRIT—A spar projecting forward from the bow of the ship.
BRACE ABACK, TO—To turn a yard around for the contrary tack.
BRACE BY, TO—To turn the yards so some sails are aback and some drawing.
BRACE IN, TO—To turn a yard more thwartwise.
BRACE PENNANT—Ropes securing the brace blocks to the yard arms.
BRACES—Ropes fastened to the yard arms to turn them about.
BRACE UP, TO—To turn a yard nearer the direction of the keel.
BRAILS—Ropes to draw up the foot, leech, and other parts of fore-and-aft sails for furling or when tacking.
BRAIL IN, TO—To haul down a spanker.
BRAIL UP, TO—To haul up a spanker, for furling.
BREACH, TO—Said of whales when leaping out of the water.
BREAK OUT, TO—To fetch stores out of the hold.
BREAK WATER, TO—When whales or fish show themselves out of water.
BRIDLES—Ropes attaching the leeches of square sails to the bowlines.
BRING TO, TO—To make a ship stationary, by bracing some sails aback and keeping others full, to counterpoise each other.
BRING UP, TO—To come to an anchor.
BROACH TO, TO—To veer suddenly into the wind so as to lay the sails aback, in heavy weather.

APPENDIX 305

BUG-LIGHT—An iron basket used as a lantern, burning blubber scraps.
BUNT—The middle part of a square sail.
BUNTLINES—Ropes attached to the foot of a square sail, to haul it up.
BY THE BOARD—Over the side.
BY THE HEAD—When a ship is deeper in the water forward than aft.

CABLE, TO PAY OUT THE—To pass it out of the hawse hole.
CABLE, TO SERVE THE—To wrap the cable with anything to keep it from chafing.
CABLE, TO SLIP THE—To let it run clear out and overboard.
CAP—A block of wood which secures the topmast to the lower mast.
CAPSTAN—A winch used for hoisting the anchor or making heavy pulls.
CAREEN, TO—To heave a vessel down on one side in order to clean or repair the bottom, also to lie or heel over.
CARRY AWAY, TO—To break, or lose.
CASE, THE—Of a whale, the upper part of the head.
CASE BUCKETS—Buckets used to bail spermaceti from a whale's head.
CAT-BLOCK—A large block used for drawing the anchor up to the cathead.
CATHEAD—Timber or crane extending over the bow to draw up the anchor clear of the ship's side.
CHAFING-GEAR—Windings and bindings of any material to prevent chafing of sails or rigging.
CLAP ON, TO—To make fast or move rapidly.
CLEATS—Blocks of wood to fasten ropes to.
CLEWS—The principal corners of sails to which the sheets are attached, e.g., for square sails, both lower corners.
CLEW DOWN, TO—To haul the yards down on the cap by manning the clew lines.
CLEW UP, TO—To haul up for furling.
CLEW LINES—Ropes from the yards to the lower corners of the sails by which the clews are hauled up.
CLOSE-HAULED—Sailing with the sails full and as near the wind as possible.
CLOSE-REEFED—Sails reduced in area as much as possible, by reefing.
COAMINGS—The raised borders of the hatchways.
COURSES, THE—The lowest square sails of a square-rigged ship. To furl the courses is to furl the mainsail and foresail.
CRANE—A davit used for hoisting the boats.
CUT IN, TO—To cut off the blubber from a whale.
CUTTING GEAR—The tools used in cutting the blubber from a whale.
CUTTING STAGE—A platform suspended over the side of a whaleship from which whales are cut in.

DART, TO—To use the harpoon or lance at whales.
DART GUN—Similar to bomb gun except that a harpoon instead of a bomb is fired from it.

Davit—A crane used for hoisting the boats.
Deck-tub—A large tub on the deck of a whaling ship.
Dhow—A lateen-rigged vessel of the Indian Ocean.
Dog watch—The watches just before and after supper on a whaler.
Doldrums—An area near the equator between the trade winds where there is little or no wind.
Double reef, to—To take in two reefs of a sail.
Downhaul—A rope passing along a stay and through the hanks of staysail or jib, made fast to the upper corner of the sail to pull it down.
Drug, to—To fasten to the end of the whale line.
Duff—Dough or paste, also a boiled pudding.
Dugout—A hollowed-out tree trunk used as a canoe.
Dunnage—Wood laid at bottom of a ship to keep the cargo dry.

Ease the ship—The command to put the helm hard alee so as to meet the waves bow on, when close-hauled and likely to plunge heavily.

Fairway—The channel of a bay or harbor.
Falls—Another name for tackles, usually applied to those at the davits.
Fall off, to—A ship falls off when she will not hold as close to the wind as she ought.
Fast—Fixed, secure, e.g., a fast whale is one attached to a whale line.
Fathom—Six linear feet.
Felucca—A fast lateen-rigged ship used in the Mediterranean.
Fill, to—To brace the yards so that the wind may strike the after surface of the sails.
Flog a cask, to—To tighten or drive up the hoops.
Flukes—The tail of a whale or the palms of an anchor.
Fluke-chains—Chains passed around the small of a whale in front of its flukes to secure or hoist it.
Fluky—Said of wind when gusty or puffy.
Flurry—The death struggle of a whale.
Foot-ropes—Ropes running under the yards for the men to stand on when furling sail.
Fore—That part of a ship nearest the bow.
Fore and aft—The lengthway of a ship.
Forecastle—The forward cabin of a vessel where the sailors live; on a whaler, usually between decks.
Foremast—The mast nearest the bow.
Foul, to—To tangle, twist or be caught, e.g., fouled lines are whale lines from several boats, all tangled.
Freshen, to—To increase, said of wind growing stronger.
Full and by—Same as close-hauled.
Furl, to—To wrap or roll a sail on the yard.

APPENDIX

GAFF—The upper spar for a spanker on the mizzen mast, or a strong hook on a handle for handling blubber.
GALLEY—The cook room on a ship.
GALLIED—Said of a whale when frightened.
GAM—A social visit between sailors at sea.
GASKET—A piece of flat rope or plait to fasten the sails to the yard.
GEAR—The equipment used for any operation, e.g., boat gear is the oars, sails, and equipment for a boat.
GIG—A long, narrow, fast, clinker-built rowboat.
GONIES—Albatrosses.
GOOSE-WINGED—Said of the outer extremities of a square sail when loose, the rest being furled.
GRIPES—Ropes holding down the boats after they are hoisted into place.
GROMMET—A ring worked into the sail for an eyelet hole.
GUNWALE—The upper rail of a vessel's or boat's side.
GURRIE—The greasy mess on a whaler's deck after cutting in and trying out a whale.
GUY—A rope to steady a boom.
GUY OUT, TO—To set the guys for a boom.

HALYARD—A rope or tackle for hoisting and lowering sails.
HARD ALEE—The helm thrown way over to leeward.
HARPOON—A barbed spear thrown at whales with whale line attached.
HATCH—The covering of an opening in the deck.
HAUL ABACK, TO—To slacken the advance of the ship.
HAUL LINE, TO—To pull on the whale lines to draw nearer to the whale.
HAUL HOME, TO—To pull the clew of a sail as far as it will go.
HAUL UP THE COURSES, TO—To haul up the mainsail and foresail.
HAWSER—A small cable.
HEAVE, TO—To turn about a windlass or capstan, or to lie to.
HEAVE THE LEAD, TO—To throw the lead overboard to take the depth of water.
HEAVE THE LOG, TO—To throw the log overboard to ascertain the speed of the ship.
HEEL, TO—To incline to one side.
HELM, THE—Same as the tiller; a steering arm.
HOLD, THE—The lower part of the ship used for storage. On a whaler there are the fore, main, and after holds.
HOLD, TO STOW THE—To store things away in the hold.
HORSE PIECES—Pieces of whale blubber about two feet long cut from the blanket pieces, ready for mincing.
HOVE-TO—Said of a ship when her headway is stopped, by bringing her up in the wind.

IN IRONS, TO PLACE—To imprison and shackle a man for misdemeanors.
IRONS—A general term for harpoons and lances used in whaling.

JETTY—A landing wharf or pier.
JIB—A triangular sail extending from the head of the foremast to the bowsprit.
JIB-BOOM—A spar that runs out on the bowsprit.
JIGGER WHEEL—A trinket made of ivory by whalers.
JUNK—Pieces of old cable from which mats and gaskets are made, also a Chinese sailing ship.
JUNK OF A WHALE—That part of the head of a whale under the case.
KEDGE—A small anchor with an iron stock.
KEG, BOAT—Water keg in a whaleboat.
KEG, LANTERN—Keg for food, matches, candles, etc. in a whaleboat.
KNOT—A division in the log line, the number of knots passing in 28 seconds representing the number of nautical miles per hour. A speed of eight knots is the same as eight nautical miles per hour. A nautical mile is 6,080 feet or about 1 1/6 land miles.
LABOR, TO—To pitch and roll heavily.
LANCE—A sharp spear used to kill whales.
LANYARDS—Of the shrouds, the small ropes at the ends of the shrouds by which they are pulled tight.
LARBOARD—The left or port side of a ship.
LAY, THE—The share of the profit of one of a whaler's crew in the catch.
LEE, THE—The side sheltered from the wind.
LEE SHORE—A shore on the lee side of a vessel.
LEE SIDE—The side furthest from the point of wind.
LEECHES—The borders or edges of a sail.
LEECH LINES—Ropes fastened to the middle of the leeches of the mainsail and foresail serving to truss these sails up to the yard.
LEECH-ROPE—That part of the bolt rope to which the edge of a sail is sewed.
LEEWARD—Toward that part of the horizon to which the wind blows.
LEEWAY—The lateral movement of a ship to leeward.
LIFTS—The ropes to the ends of the yards suspending them from the mastheads.
LIGHTER—A barge for transferring stores, etc., between shore and ships.
LIGHT SAIL—Any sail used only in light winds.
LOBTAIL—Said of a whale lashing the water with its tail.
LOG—A device for measuring the rate of a ship's speed.
LOG-BOOK—The book in which the record of the ship's voyage is kept.
LOG-SLATE—A slate on which notes are kept to be recorded later in the log-book.
LOGGER-HEAD—An upright piece of timber in a whaleboat around which a turn of the line is taken to prevent the line running out too fast.
LOOSE, TO—To unfurl or cast loose any sail.
LOOSE WHALE, A—One to which no line is attached.

APPENDIX

LUBBER—A sailor who does not know his duty.
LUFF, TO—To turn the head of a vessel towards the wind.
LUFF-TACKLE—A large tackle consisting of a double and single block.
LUG-SAIL—A four-sided sail bent upon a yard which hangs obliquely to the mast at one-third of its length.
LYING OFF AND ON—Sailing back and forth across the entrance to the harbor.
LYING HOVE-TO—Ship lying floating, not advancing.

MAIN TOP—The platform at the head of the mainmast.
MAKE SAIL, TO—To increase the quantity of sail set on a ship.
MAN THE WINDLASS, TO—To send men to place the poles in the windlass ready to turn.
MAN THE YARDS, TO—To send men up on them.
MARLINE—Small tarred twine used for wrapping or tying.
MARLINESPIKE—A pointed iron tool for separating the strands of a rope when splicing.
MARTINGALE—The lower stay of the jib-boom.
MAST, FORE—The mast nearest the bow.
MAST, MAIN—The center mast of three.
MAST, MIZZEN—The mast nearest the stern.
MAST CLOTH—The lining in the middle of the after side of topsails, etc., to prevent chafing by the mast.
MASTHEAD—The top of the lower mast, also on a whaler applied to the lookout man.
MAT—A webbing of yarn to prevent chafing.
MAT, SWORD—A mat made of woven ropes to prevent chafing.
MAT, THRUM—A mat made of tufted yarn to prevent chafing.
MILL, TO—To move in circles or to continually change direction, e.g., referred to as of whales milling.
MINCING KNIFE—A knife used for mincing when feeding blubber to mincing machine.
MINCING MACHINE—A machine used for cutting blubber fine.
MONSOON—A periodic wind in the Indian Ocean blowing from S.W. from May to September and N.E. from October to December.

NIPPERS—An attachment to the harpoon for attaching the line.
NIPPERS, TO DRUG THE—To fasten the whale lines to the harpoon attachment.

OAKUM—Untwisted rope or rope-fiber impregnated with tar and used for caulking.
OFF AND ON—Leaving the land on one tack and coming near it on the other.
ON THE BEAM—At right angles with the keel.
ON THE BOW—An arc including 45° from the bow.
ON THE QUARTER—An arc including 45° from the stern.

OVERHAUL, TO—To haul a fall of rope through a block until it is slack, also to overtake.

PAINTER—A rope by which a boat is made fast.
PARCEL, TO—To cover with slips of canvas and tar same.
PARCEL A ROPE, TO—To put old canvas on a rope before the serving is put on.
PARCEL A SEAM, TO—To lay a strip of canvas over a joint or seam after it is caulked—applied to wooden ships.
PARCELING—Slips of canvas, spirally bound about a footrope, previous to being served.
PAWL, A—A ratchet to prevent a windlass from rolling back.
PAY OFF, TO—To make a ship's head recede from the wind by backing the head sails.
PAY OUT, TO—To slacken or let a rope run out.
PEAK—The angle formed by a gaff and mast of a fore-and-aft sail, also the upper end of the gaff.
PINS—Same as belaying pins.
PITCH, TO—To plunge fore and aft alternately.
PLACE IN IRONS, TO—To handcuff and keep in prison quarters.
PLIMSOLL MARK—The load-line mark of British vessels to show the depth of submergence.
POINTS, THE—The points of the compass.
PORT, TO—To go to the left—now used altogether in place of larboard.
POTS—Large iron pots for boiling out the blubber.

QUADRANT—An instrument used for measuring altitudes, now superseded by the sextant.
QUARTERDECK—That part of the upper deck abaft the mainmast, reserved for officers.
QUARTERMASTER—A petty officer who attends to the helm, signals, binnacle, etc.

RANGE OF CABLE—A length of cable sufficient to anchor the ship.
RATLINES—The small horizontal ropes on the shrouds by which the men go aloft.
REEF—Any single reef for square sails; also a shoal barrier, a coral reef.
REEF, TO—To reduce the area of a sail by taking in with the reef points.
REEF BANDS—Pieces of canvas sewed across a sail to strengthen it at the eyelet holes of the reefs.
REEF LINES—Small lines, stretched across the reefs and spliced into the cringles, for men to catch hold of.
REEF POINTS—Plaited ropes made fast to sails for the purpose of reefing.
REEF TACKLES—Tackles used to pull the skirts of topsails close to the yards to ease the labor of reefing.
REEVE, TO—To put a rope through a block.

APPENDIX

RENDER, TO—To yield to the efforts of mechanical power or to slack off. In whaling, to melt down blubber into oil.
RIDING LIGHT—A white light hung at masthead or in the rigging when ship is at anchor.
RIGGING, THE RUNNING—The ropes usually running through blocks that handle the spars and sails.
RIGGING, THE STANDING—The permanent cables, ropes, etc. that hold the masts and spars in place.
RIGHT THE HELM, TO—To place the helm parallel to the keel.
RINGS, THE—An iron cage at masthead for the lookout.
RISING, A—The reappearance of a whale after diving.
ROADSTEAD—A protected place where ships may ride at anchor.
ROPE-BANDS—Short flat pieces of plaited rope having an eye at one end used to tie the upper edges of square sails to the yards.
ROVINGS—Unraveled rope fibers.
ROWLOCK—The fulcrum for an oar when rowing.

SCARFING THE BLUBBER—To cut the blubber of a whale in channeled strips, ready to haul by the hook.
SCRAPS—The solid remainder of the blubber after trying out the oil.
SCRIMSHAW WORK—Ornaments, trinkets, carvings, etc., made by sailors in spare time.
SCUPPERS—Openings in the side of a vessel to let water run off the deck, also the deck waterways of a ship.
SCUTTLE BUTT—A large cask for holding drinking water.
SEA CHEST—The personal chest of a sailor for his belongings.
SEA ROOM—Sufficient clearance from a dangerous coast to be safe.
SEAM—Two edges of canvas where sewed together, also the caulked timber joints of a wooden ship.
SEIZE, TO—To make fast or bind.
SEIZING—The operation of binding two ropes together with cord or spun yarn, also the binding material itself.
SENNIT—A small plaited rope made from spun yarn.
SERVE, TO—To wind anything around a rope to prevent chafing.
SERVINGS—Spun yarn or canvas used to protect rope from chafing.
SET THE WATCH, TO—To place the men in their proper positions for their turn on watch.
SETTING UP IRONS—Inserting wooden shafts into the whaling lances.
SEXTANT—An instrument used to determine the altitude of the sun from the horizon, from which the ship's position is calculated.
SHANK—The shaft of an anchor, lances, etc.
SHEER, THE—The side curve of a ship between bow and stern.
SHEER, TO—To deviate from a course.
SHEER OFF, TO—To turn aside.

SHEER POLES—Poles used in hoisting or bracing.
SHEET—A rope fastened to the lower corner of a sail to hold it in position.
SHEET HOME, TO—To haul the sheets home to the block on a yard arm.
SHIP TRACK—A much-traveled ship route.
SHIPSHAPE—In an orderly, seamanlike manner.
SHIVER, TO—To make the sails shake.
SHOOKS—Unassembled parts of barrels and casks.
SHROUDS—A range of large ropes from mastheads to sides of ships to support the masts.
SIDE LIGHTS—The port and starboard sailing lights.
SIGHT, TO TAKE A—To take an observation on the sun with sextant or quadrant.
SLOP CHEST—On a whaler, small stores available but charged to the men.
SMALL OF A WHALE, THE—That portion of a whale of smallest cross-section just ahead of the flukes.
SMOKE-SAIL—A sail erected over the try pots to cause the smoke to rise from the deck.
SOUND, TO—Said of a whale when diving towards the bottom.
SOUNDINGS, TO TAKE—To ascertain the depth of water with a leaden plummet.
SPADE—A cutting tool used to cut and strip blubber from a whale.
SPANKER—A fore-and-aft sail with boom and gaff on mizzen mast.
SPARS—A general term for masts, yards, booms, etc.
SPLICE, TO—To join two ropes together by uniting the strands.
SPRING LINE—A diagonal line from bow or stern to a point on a wharf.
SPRIT-SAIL—A sail extended by a sprit or pole crossing a fore-and-aft sail.
SPUN YARN—Two or three rope yarns twisted together.
STANCHIONS—Any upright support on a ship's deck.
STAND IN, TO—To head toward the shore.
STAND ON, TO—To keep in the course.
STARBOARD—The right-hand side.
STAYS—Large ropes extending forward from the mastheads to the deck.
STAYS, IN—The act of going from one tack to the other.
STAYSAIL—Any sail extended on a stay.
STEERAGE—The space below the deck nearest the rudder.
STEERAGE WAY—Sufficient movement of a ship to make her answer the helm.
STEP A MAST, TO—To fix the foot of a mast in place and erect it.
STEP POLES IN THE IRONS, TO—To fix wooden shafts to the whaling irons.
STERN—The after end of a vessel.
STEVEDORE—A man who loads or unloads vessels in port.
STIRRUP—Rope secured to a yard with a ring in its lower end for supporting a footrope.
STOPS—Small lines used to bind or secure something.
STOVE BOAT—A damaged or broken boat.

APPENDIX

STOWAWAY—One who conceals himself on a vessel to obtain a free passage.
STRIKE—To beat against the bottom or a shoal; to harpoon a whale.
STUDDING SAIL OR STUNSAIL—Sails extended beyond the leeches of the principal sails, appearing as wings to the yard arms.
SWEEPS—Long oars used to steer or propel boats.
SWIFTER—A line used to retain the bars of a capstan in their sockets.
SWIFTER, TO—To tauten.
SWORD MATS—Mats to prevent chafing.

TACK, TO—To turn a ship by the sails and rudder when sailing against the wind.
TACKS—The lower and forward corners of all fore-and-aft sails, also the lower and forward corners of all courses.
TACKLES—An assemblage of ropes and pulleys for hoisting and pulling.
TAFFRAIL—The rail around a ship's stern.
TAKE THE SUN, TO—To make an observation on the sun with sextant or quadrant to figure the ship's location.
THROAT—The upper corner of a gaff sail next the mast.
THRUM MATS—Mats of spun yarn or rope applied to rigging to prevent chafing.
THWARTS—A rower's seat reaching athwart a boat.
TIER—The row or rank of casks in the hold of a whaling ship.
TILLER—The lever of a rudder.
TRADES OR TRADE WINDS—Winds blowing from N.E. to S.W. on north side of equator, S.E. to N.W. on the south side of equator with doldrums between.
TRYING—Lying-to in a gale of wind under small sail; also boiling out oil from the whale blubber.
TRY POTS—Iron pots used in trying out whale blubber.
TRYSAIL—A small three-cornered sail used in heavy weather.
TRYWORKS—The equipment used in boiling oil.

VANE—A small flag flown at each masthead.
VANGS—Traces or ropes to steady the mizzen gaff, extending down to the after deck.
VEER, TO—To change direction.
VISE BENCH—A work bench with a vise on it.

WAIF—A flag used to mark a whale.
WAIST OF A SHIP, THE—That part of a ship between the foremast and mainmast.
WARP, THE BOAT—The mooring rope or painter of a boat.
WATCH, THE—One half the crew, either the port or starboard watch, an allotted time (four hours except the dog watches) for being on deck for duty.

Watch tackle—A two-sheave tackle used to handle yard arms.

Wear, to—To turn a ship around by the stern to catch the wind as opposed to tacking.

Weather beam, the—On the weather side, at right angles to the keel.

Weather braces, the—The braces on the weather side.

Weather helm, to have a—Said of a ship inclined to come too near the wind.

Whaleboat—An open, double-ended boat, usually about 30′ x 6′, manned by mate, boatsteerer and four oarsmen.

White horse—Tough, sinewy substance in the head of sperm whales just above the upper jaw.

Whitewater—Said of whales splashing, making the water appear white.

Windlass—A machine for hauling or hoisting, a capstan.

Windward—Toward the wind.

Yard arm—Either end of a yard on a square-rigged vessel.

Yards—The spars on which the sails are spread.

Yawing—The motion of a ship when she deviates from her course to the right or left.

THE "LAY," CATCH, OR WAGES

Wages, strictly speaking, were not paid on whaling ships. Instead, the officers and men were paid according to the catch or "lay" and that, to some extent, was regulated by the size of the ship.

According to Captain Ferguson, the lay was as follows:

The captain	received 1/12	of the catch
The 1st mate	received 1/18	of the catch
The 2nd mate	received 1/28	of the catch
The 3rd mate	received 1/35	of the catch
The 4th mate	received 1/60	of the catch
The cooper	received 1/60	of the catch
Each harpooner	received 1/80	of the catch
The steward	received 1/90	of the catch
The cook	received 1/110	of the catch
Each able seaman	received 1/150	of the catch
Each green hand	received 1/175	of the catch
The cabin boy	received 1/200	of the catch

The catch included all oil, whalebone, ivory, ambergris, etc., taken on the voyage, and was divided in the proportions indicated. These proportions were not standard and varied considerably on other ships.

On this voyage, the *Kathleen* took few right whales, consequently the whalebone was only a small part of the catch. Many humpback whales were taken but their bone was valueless. No ambergris is mentioned in the diary as having been taken in all the four years.

On this voyage of the *Kathleen*, the total amount of oil taken was as follows:

9/13/80 Sent 416 bbls. home by barque *Veronica* from Fayal, Azores.
3/21/81 Sent 460 bbls. home by barque *Chas. W. Morgan* from St. Helena.
3/25/82 Sent 982 bbls. home by schooner *Lottie Beard* from St. Helena.
10/26/82 Sent 577 bbls. home by schooner *Lottie Beard* from St. Helena.
3/31/83 Sent 464 bbls. home by schooner *Lottie Beard* from St. Helena.
5/22/84 Took 1801 bbls. home by barque *Kathleen* to New Bedford.

        4700 bbls. @ 31.5 gals per barrel totals 148,050 gals. for voyage.

The value in dollars and cents for whale oil depended not only upon its quality, which varied for each kind of whale, but upon its market value upon arrival.

The very finest oil, which was porpoise-jaw oil, the best of all lubricants for watches, would fetch as much as five dollars per gallon. The quality and price of oil from sperm whales varied according to whether it was head oil, spermaceti, body oil, or fluke oil. The oil of different whales, right whales, humpbacks, blackfish, etc., was of varying quality and brought different prices.

In the seventies, during the earlier days of Captain Ferguson's whaling, oil brought as much as $1.20 per gallon. In the latter days of his experience, when the demand for oil was falling off, the price fell to $.60 and possibly $.40 per gallon. At the latter price, whaling as carried on by the small whaling barques could not be made to pay.

If we assume that the 148,050 gallons of oil taken by the *Kathleen* on this four-year voyage averaged $.70 a gallon, its total value would be $103,635.00. A harpooner's share @ 1/80 would total a little less than $1,300.00 for the voyage or about $325.00 per year, from which would be deducted any money drawn or advanced on the voyage, and the cost of his calls on the slop chest for clothes, tobacco, etc. Added to this would be his share of the bone, ivory, and ambergris, plus bounty money.

Should any member of the crew run away or desert, he would lose his "lay" or share of the catch. To prevent desertion, and to avoid recruiting during the voyage, all whalers avoided busy ports and those where there was a large population. When it was necessary to go into a port for water, wood, or to ship oil home, it was usually at such an out-of-the-way place that it would be safe to let the crew take a run on shore and be reasonably sure that they would all return to the ship. A whaler could not operate successfully short-handed. The number of men composing a whaler's crew was usually three times greater than on a merchant ship, which might also be a much larger boat. At the above-mentioned ports, additional men for replacing any of the crew who had run away would be difficult to obtain.

On the whole, men on a whaler lived better and had better food and wages than men on a packet ship. For comparison, the wages on packet ships during the same period were:

Seamen	from $12.00 to $15.00 per month
2nd mates	from $30.00 to $35.00 per month
1st mates	from $35.00 to $45.00 per month

It will be seen, therefore, that a harpooner's earnings on the *Kathleen* compared favorably with those of a second mate on the packet ships.

A job on a whaler may have been dirty, greasy, and malodorous, but the thrills of the chase and shore leave at interesting out-of-the-way places surely made it attractive.

www.ingramcontent.com/pod-product-compliance
Lightning Source LLC
Chambersburg PA
CBHW071109060525
26252CB00034B/378